FALSE MESSIAH

FALSE MESSIAH

DARWINISM AS
THE GOD THAT FAILED

NEIL THOMAS

SEATTLE   DISCOVERY INSTITUTE PRESS   2025

Description

Ever since *The Origin of Species* appeared in 1859, Charles Darwin's followers have co-opted him as the patron saint of materialism. In *False Messiah*, longtime Darwinist and agnostic Neil Thomas looks at how Darwinism triumphed, and he challenges the official story with an exploration richly informed by his expertise in rhetoric and cultural history. What of Darwinism's present status? Like Marxism—that other great materialist theory of the Victorian age—Darwinism has become "the God that failed." As Thomas shows, although Darwinism has done incalculable damage to our culture, there is hope for renewal in contemporary scientific findings that are reinvigorating the argument from design as well as in what amounted to an ingenious flanking movement against modern materialism—the spiritually charged philosophy of nature developed in the poetry of William Wordsworth.

Copyright Notice

Library Cataloging Data

False Messiah by Neil Thomas

Cover design by Tri Widyatmaka

212 pages, 6 x 9 inches

Library of Congress Control Number: 2025939267

ISBN: 978-1-63712-080-4 (paperback), 978-1-63712-082-8 (Kindle), 978-1-63712-081-1 (EPUB)

BISAC: SCI027000 SCIENCE / Life Sciences / Evolution

BISAC: SCI034000 SCIENCE / History

BISAC: SCI075000 SCIENCE / Philosophy & Social Aspects

Publisher Information

Discovery Institute Press, 208 Columbia Street, Seattle, WA 98104

Internet: discovery.press

Published in the United States of America on acid-free paper.

First Edition, July 2025

ADVANCE PRAISE

In this fascinating and wide-ranging book, Neil Thomas draws on his expertise in European intellectual history and linguistics to examine both the content and context of Darwin's writings. He draws in detail on the words of Darwin's contemporaries, such as Charles Lyell and Asa Gray, to examine how Darwin's ideas were received in the nineteenth century. He also considers the literary context, in the world of William Wordsworth, George Eliot, and Algernon Charles Swinburne. There is a particularly interesting section on Alfred Russel Wallace, the co-discoverer of Darwin's evolutionary theory, who later changed his mind on some of the key aspects.

Thomas also looks carefully at some of the details of Darwin's text, and the difficulties of translating the concept of natural selection into other languages, and sets out what these reveal about Darwin's lack of conceptual clarity. I found the discussion of Darwin's religious confusion and struggles, and the claim set out here that Darwin's presentation of natural selection comes close to that of a goddess figure, to be thought-provoking and insightful.

The book provides a stimulating challenge to the lazy assumption that Darwin's work provided a simple and satisfying scientific explanation for all aspects of the biological world. Anyone who holds onto that view after studying the issues discussed here really is clinging to a false messiah.

—**Peter Jeavons**, Senior Research Fellow, St Anne's College, Oxford; Emeritus Professor, University of Oxford

How in the world could a theory as transparently lame as Darwin's win over most late-nineteenth-century academics? Neil Thomas's brilliant new book, *False Messiah: Darwinism as the God That Failed*, shows that many intellectuals then were already hoping to explain away the obvious designedness of life, in order to subsume biology into a totalizing, human-centered Enlightenment vision. Darwin merely offered them what they were hoping for. Ironically, as Thomas notes, it is the very progress of modern science—especially the discoveries of non-mechanistic quantum reality and the information-suffused molecular layers of life—that has ultimately quashed their efforts, whether or not they yet realize it.

—**Michael J. Behe**, Lehigh University Professor of Biological
    Sciences and author of *Darwin's Black Box*, *The Edge of Evolution*,
    and *Darwin Devolves*

A beautifully written exploration of Darwin's thinking, set against the historical context of his time and the diverse responses to his controversial ideas. Neil Thomas offers fascinating insight into the reception of *The Origin of Species* and the deep uncertainties Darwin himself harbored about his collected ideas. This work showcases careful scholarship and compelling analysis, exposing the flimsy foundations of the many persistent and popular misconceptions surrounding Darwin's views.

—**David J. Galloway**, former President of the Royal College
    of Physicians and Surgeons of Glasgow; Consultant Surgeon
    and Honorary Professor of Surgery, University of Glasgow,
    University of Malaysia, and the Chinese University of Hong
    Kong; author of *Design Dissected* and *Controlled Chaos*

Readers of *False Messiah* will probably already know that Charles Darwin was a product of Victorian culture—in most respects, a typical one. Neil Thomas sheds useful light on the open-minded character of that culture, which allowed *The Origin of Species* to receive the critically sympathetic hearing that it did, even though the book challenged many of the orthodoxies of the time. As made abundantly clear in these pages, a key factor was the absence of a monopoly over scientific wisdom. Learned people from all fields were given a voice in such existential matters as the origins and meaning of life. Now, nearly two centuries later, various crises of scientific authority are returning us to a similar sensibility. But whereas Victorian open-mindedness opened the door to Darwin's theory of evolution, today's open-mindedness may soon see it to its departure.

—**Steve Fuller**, Professor of Social Epistemology, University of Warwick, UK, and author of nearly thirty books, including *Science vs Religion?* and *Dissent over Descent*

"Scientists animated by the purpose of proving themselves purpose-less make an interesting subject for study," the philosopher A. N. Whitehead wrote wittily and wisely several decades ago. Prof. Neil Thomas's deft new book is as timely, judicious, and profound a brief account as could be desired of the defects, dangers, and enduring damage of orthodox Darwinism.

False Messiah contains a tight argumentative line and extra-ordinarily profound insights and felicitous locutions, as in documenting in detail the "stubborn disparity between Darwin's theory and the uncooperative facts of objective reality." Its depiction and rehabilita-tion of Darwin's noble contemporary, the pioneering naturalist Alfred Russel Wallace, a repentant reductionist, is particularly illuminating and heartening.

Thomas's articulate discussion of the protest against reductionism by the poet William Wordsworth (an instantiation of the Romantic and philosophical protest also found in Blake, Coleridge, Carlyle, and Dickens) reminds one of the truth and importance of Lionel Trill-ing's point made over a half-century ago: "Wordsworth's great auto-biographical poem *The Prelude* gives the classic account of the damage done to the mind... by the scientistic conception of mind that pre-vailed among intellectuals at the time of the French Revolution." That scientistic conception retains great influential force, damaging human mental health and prospects for truth and justice in the twenty-first century. Thomas's lucidly written book is an antidote to it.

—**M. D. Aeschliman**, Professor Emeritus of Education, Boston University and author of *The Restoration of Man: C. S. Lewis and the Continuing Case Against Scientism*

Although I am a theologian and not a scientist, nonetheless I warmly commend this unusual book by Neil Thomas. Interestingly, Darwin himself was not always totally convinced by his theory of natural selection. And as Neil Thomas pointed out in his previous book, *Taking Leave of Darwin*, modern evolutionary theory rests on a foundation considerably less stable and reassuring than rock. In *False Messiah*, Thomas shows that Darwin's theory "cannot bear the weight of its role as Gospel for a brave new age of atheistical enlightenment."

—**Paul Beasley-Murray**, former Principle of Spurgeon's College, London, and author of many books, including *There Is Hope*

Darwinism is demoralizing. The more you believe Darwin provides an explanation for life, the more meaningless that life becomes. Darwin gave atheists false hope that their decision to reject purpose in the universe is right, and that living without ultimate hope is good. As Neil Thomas expertly demonstrates, Darwin benefited from a rising tide of secularism that warmly embraced his material explanation for life. Thomas's book may well benefit from a rising tide of skepticism in the sufficiency of that materialist account, as modern science unravels the Darwinian story. This book gives good reasons to see through the messianic pretensions of Darwinism, and to follow instead the divergent path of Darwin's peer and co-discoverer, Alfred Russel Wallace, back towards the overruling intelligence of God.

—**Alistair J. McKitterick**, Senior Lecturer at London School of Theology, Consulting Fellow for the Whewell Centre, Cambridge, and author of *Faithful Science*

Rarely have I encountered a book so stimulating, yet such a joy to read. With superb insight and his delightful prose, Neil Thomas invites us to come along on his own journey of discovery. Building on his deep expertise in the history of thought, Thomas presents a profound new assessment of the ideas that shape our culture and our lives. An essential resource for anyone wrestling with the issues of origins and their impact on the meaning of life.

—**Steve Laufmann**, co-author of *Your Designed Body* and chair of the Conference on Engineering in Living Systems

Neil Thomas tackles with considerable verve and wit the central conundrum of our age: the ever growing hiatus between what we think (or thought) we knew about the practicalities of genetic inheritance and its inscrutable complexities as revealed by modern science. Specifically, despite prodigious endeavor, the empirical evidence for the mechanisms of evolutionary transformation has proved elusive. His interpretation of how nonetheless it has dominated biological thought for 160 years, providing an all-encompassing explanation for the near infinite diversity of form and attributes of the living world, is as lucid as it is persuasive.

—**James Le Fanu**, Fellow of the Royal College of Physicians, winner of the *Los Angeles Times* Book Prize, and author of *Why Us?* and *The Rise and Fall of Modern Medicine*

CONTENTS

PROLOGUE

I N A PREVIOUS VOLUME, *TAKING LEAVE OF DARWIN*, I TELL WHY I WENT from a member in good standing of the Darwinian establishment to a person surprised to discover that modern evolutionary theory rests on a foundation considerably less stable and reassuring than rock. Specifically, on closer examination of the theory I was confronted with the bewildering spectacle of some indisputably clever people making unsupported claims in favor of a hypothesis originally advanced upon a surprisingly rickety foundation.

I found that when Charles Darwin, together with later biologists as well as cosmologists, would reference purely notional processes such as "natural selection," "self-assembly," or "a cosmic imperative," these terms amounted to little more than a side-stepping of a host of imponderables. Hence the disquiet I felt at discovering that all their scientific-sounding livery were but the regalia of so many empty referents carried along by magical thinking. And I mean "empty referents" in its formal linguistic sense—words without demonstrable content within, or ascertainable relation to, the real world.

Moreover, I found that Darwin and his successors had involuntarily, by the sheer incredibility of their claims, furnished presumptive evidence for the role of a special dispensation behind the origin and evolution of life on earth, a dispensation far beyond the parameters of a purely naturalistic analysis.[1]

The present volume continues my intellectual journey, for it soon became clear to me that there were many further avenues still needing to be explored. In simplified terms, the present work takes relatively short sections of Chapters 2 and 3 of *Taking Leave of Darwin* and expands them into an entire book.

In that previous work, I briefly surveyed the early and later voices of dissent from Darwin and weighed the back and forth between Darwin and his defenders on the one side and his more thoughtful critics on the other. But this is a rich vein, and I barely scratched the surface. Here, pick in hand, we will delve deeper.

I should make clear from the outset that that my cheeky title, *False Messiah*, is not meant to describe Darwin as he ever styled himself. He actively resisted being co-opted as the figurehead of a secularist religion. In part this may have been the respectable Victorian gentleman steering clear of anything redolent of French radicalism, but the effort also seems to have stemmed from a genuine personal ambivalence concerning matters metaphysical.

Certainly, much of his professional work was devoted to substantiating the anti-theistic convictions of his grandfather, Erasmus Darwin, but Charles was much more prone to vacillate over the broader implications of his theory, brow knit, than to proclaim, fist in palm, some particular materialistic religion. Moreover, there is compelling evidence, analyzed in this book, that in later years he developed doubts about his formula for "creation without a creator." In mining his personal correspondence and autobiography, one finds that in older age he became almost as much a natural theologian as a natural scientist.

However, none of that has prevented a body of Darwin's followers from co-opting him as a sort of latter-day messiah for a materialist program of cultural and civilizational transformation. I will argue that employing Darwin in this way is false twice over: First, it is not what Darwin made himself out to be. Second, and as accumulating evidence has made plain, his theory—with or without the modern synthesis, and with or without the various third-way epicycles tacked onto it in recent decades to *save the appearances*—cannot bear the weight of its role as Gospel for a brave new age of atheistical enlightenment.

However one chooses to characterize the aging gentleman of Down House, his doubts about the materialistic project have been amply vindicated by advances in biology in the last seven decades—one of the topics explored in the present book. Which brings us to the present volume's subtitle. Students of the Cold War may recognize it as the title of a 1949 collection of essays by six prominent writers of the day on their conversion to, and eventual disillusionment with, communism.

My borrowing of the phrase, and thereby connecting Marxism to Darwinism, is not incidental. Although *The Communist Manifesto* was published a decade before Darwin's theory of evolution went public, Marx said he saw in Darwin's *On the Origin of Species* "the book which, in the field of natural history, provides the basis for our views."[2] Or as he put it a year later: "Darwin's work is most important and suits my purpose in that it provides a basis in natural science for the historical class struggle.... Despite all shortcomings, it is here that, for the first time, 'teleology' in natural science is not only dealt a mortal blow but its rational meaning is empirically explained."[3] In similar terms, Marx's collaborator, Friedrich Engels, enthused that Darwinism had helped to eliminate teleology from biology[4]—much as Marx, we might say, had eliminated it from social science.

Today we could invert this formulation to assert, with only slight exaggeration, that Marx has been eliminated from social science, at least Marxism as a coherent, all-encompassing political program.

Darwinism, I want to argue, is approaching a similar condition of obsolescence. There are relatively few dyed-in-the-wool Darwinists remaining. True enough, in the academy one is still asked to genu-flect before that mistiest of all dogmas, *evolution*. But among actual scientists doing serious work in biology, things have moved on from Darwin and even neo-Darwinism. As with many of our present-day social scientists, what remains in the biological sciences is a dogmatic commitment to materialism, regardless of how desperate that commit-ment may have become.

Much of what follows has benefited from my background and expertise in European intellectual history as well as in languages and linguistics. In the present volume I cast a wide net to view Darwinism against the backdrop of European cultural assumptions, with special reference to religious and theological thinking, particularly among the Darwinists. Within that context I also focus on many linguistic ambiguities and modes of literary presentation favored by Darwin, which point towards largely unexplored subtextual implications.

The volume is structured as follows. Chapter 1 describes the contemporary Darwin debate by attempting a "warts and all" sketch of the various issues at stake. In particular, I discuss how, more than 160 years on, there is no settled acceptance of modern Darwinian theory even among mainstream biologists, a reality rarely discussed in the popular press but one easily enough uncovered. To that end the opening chapter briefly surveys the more important issues which presently divide evolution experts.

Chapter 2 discusses Darwin's frequent lack of conceptual clarity—a lack which becomes evident when we look at some of his imprecise and, at times, even misleading terminology. It was this lack of clarity which drew him into the contradiction of claiming, on the one hand, that "natural selection" was a mindless process and, on the other hand, that it was a crypto-teleological process analogous to that employed by animal breeders bent on procuring the best livestock. After unmasking some ways in which Darwin's conflicted mindset could have stemmed from ways of thinking familiar to the Victorians but no longer intelligible to our present generation, I go on to document how even some modern biological researchers have been constrained by their evidence to concede that Darwin's notion of natural selection does not line up with the results of modern, empirical research in the area.

Chapter 3 considers the implications of Darwin's autodidactic education-by-installments, including his own awareness of the difficulties arising from a theory which had its origins in a stitching together of opinions he took from others rather than its emerging empirically from minutely observed fieldwork, as the Darwin mythology has it. As well as tracing modern influences on Darwin, I also consider the work of ancient atheist philosophers to whom Darwin owed a debt he was little aware of. Here too I consider the implications of Darwin's giving a new lease on life to the philosophy of atheistic atomism, cold-shouldered for almost two thousand years.

Chapter 4, "Culture Wars," begins a three-pronged analysis of the changing religious landscape of Victorian Britain. First I come at it from the perspective of writers of the period, many of whom were prescient in their ability to anticipate trends in the developing, or sometimes deteriorating, spirituality of their wider society.

Then I come to the university world of Oxford, a bellwether for intellectual developments throughout the nineteenth century and location of the famous debate on Darwinism in 1860 between Darwin's bulldog, Thomas Huxley, and Oxford's bishop, Samuel Wilberforce. Here I address Oxford's changing culture through the nineteenth century as our ancient seats of learning reluctantly adapted themselves to secularizing trends in post-Enlightenment Europe.

Thereafter I come to the grassroots, mainly the working-class movement simply styled "secularism," which showed itself in remarkable advance of Oxford in its attitude to the German Higher Criticism of the Bible. I argue that the upshot of Darwin's delaying publication of *On the Origin of Species* by almost two decades (he had already produced a pencil sketch by 1842), led to its benefiting from the rising tide of secularism which peaked in Britain in the period circa 1850–1890. The debatable logic of the *Origin* made sense within that secularizing context by dint of what is now commonly termed confirmation bias.

Later in the chapter I turn to that deuteragonist of the Darwin story, Alfred Russel Wallace, whose many scientific achievements were, in their totality, considerably more impressive than those of Darwin. I assess the reasons that eventually impelled him to forsake crucial aspects of his co-authored theory of natural selection. Wallace has often been sidelined by science historians who have described the evolutionary story in almost exclusively Darwinian terms, a tendency I resist in this volume.

Chapter 5 discusses doubts harbored by both Darwin and some of his modern legatees concerning the sufficiency of natural selection. I also chart the ways that Darwin's friends and colleagues helped him promote his ideas and yet, taking their cue from Darwin's own insecurities and hesitancies, felt no obligation to regard his ideas as sacrosanct, so much so that they offered sharply varying takes on the *Origin*. In fact, many unresolved tensions in Darwin's own mind caused him to present his *Origin* somewhat diffidently, as a theory worthy of consideration only until a better one might be advanced.

Chapter 6 is devoted to what was in effect a rival philosophy of nature developed in the poetry of William Wordsworth. I look at how the poet's considerable influence across all social classes as well as across the Atlantic Ocean stood as a philosophical barrier to people's full acceptance of Darwinism.

The seventh and final chapter considers how the recent disciplines of quantum mechanics and microbiology are difficult to square with those Newtonian notions of scientific predictability Darwin appealed to for intellectual support. This bears on the plausibility of Darwinian notions, since recent discoveries, acting as a corrective to Enlightenment presuppositions and presumptions, have undermined faith in the capacity of materialist science alone to predict outcomes and deliver truth. This, in turn, has undermined the Darwinian project itself.

1. DARWINISM AND
ITS DISCONTENTS

THE IDEA OF EVOLUTION HAS A LONG PEDIGREE IN EUROPE, FIRST among the ancient Greek atomists and energetically resurfacing in the Enlightenment. I traced that history in my previous book and won't belabor it in these pages. Here, suffice to say that what Erasmus Darwin, Charles's grandfather, called the transmutation of species (borrowing an old alchemical term), and what Charles Darwin referred to as descent with modification, was a subject which had already exercised those free-thinking eighteenth-century French writers whom history refers to as *les philosophes*.

These French Enlightenment figures toyed with the idea of the morphology of particular animal types changing over vast tracts of time. In his *L'Homme Machine* (1747), Julien Offray de la Mettrie argued that all animal forms had emerged from previous forms, so that the earthworm might be expected to transmute in time to become a considerably larger and more complex animal. Denis Diderot, anticipating conceptions later put forward by his countryman Jean-Baptiste Lamarck, mooted in his *D'Alembert's Dream* (1769) the possibility of a creature evolving through habitual functioning into another form of life altogether. Diderot even played with the bizarre (not to say physiologically illiterate) idea that those humans not required to perform manual labor might eventually become just heads.

In the midst of what others regarded as the eccentric musings of a small, self-referential coterie, it soon became clear that what was required was the identification of a causal mechanism that might render plausible the counterintuitive claim of boundless metamorphic evolution alleged by the group. The ambition was to identify what scientists from Newton onward termed a *vera causa*, a sufficient actuating means, for the changes in organismal form in the history of life.

The theorizing of Darwin's grandfather and the other transmutationists carried the idea that unaided nature, not God, had been responsible for rolling out Earth's diverse biosphere of plants and animals. Perhaps hoping to justify his grandfather's thinking on this point, Charles Darwin applied himself to finding a material mechanism underlying the evolution of all things and, hence, cut out the need for a cosmic middleman, so to speak. The idea that he (and Alfred Russel Wallace independently) hit upon he dubbed "natural selection."

The fundamentals of the idea were succinctly described by Darwin in the introduction to the *Origin*:

> As many more individuals of each species are born than can possibly survive; and as, consequently, there is a frequently recurring struggle for existence, it follows that any being, if it vary however slightly in any manner profitable to itself, under the complex and sometimes varying conditions of life, will have a better chance of surviving, and thus be naturally selected. From the strong principle of inheritance, any selected variety will tend to propagate its new and modified form.... Natural Selection almost inevitably causes much Extinction of the less improved forms of life, and induces what I have called Divergence of Character.[1]

Cynthia Russett uses the flounder fish to illustrate the idea:

> The flounder originally lacked spots, but at some point a few baby flounders were born with spots which camouflaged them when they swam over pebbles or sand and thus lessened the chance of their becoming dinner for a hungry predator. A greater number of spotted than nonspotted flounders would therefore live to become parents of a new generation—some without spots, but many spotted. Extended over vast stretches of time, this process of selection would lead to permanent modification of the flounder into a species all members of which possessed spots. And as selected individuals

might form a distinct variety, so, *with a greater effort of imagination*, one might suppose varieties being elevated into separate species.[2]

Such is the Darwinian narrative in simplified form, the story of how all life is said to have evolved naturally, without benefit of intelligent oversight, from the state of unicellular beginnings to that of highly complex plants and animals. On this accounting, natural selection is the previously unrealized means by which random variations were threshed and the surviving features and organisms put on the path from simpler to more complex, the changes leading in time towards the development of new species, families, and phyla.

I have drawn special attention to Russett's words "with a greater effort of imagination" in the quotation above because Darwin's contemporaries, as well as all present-day scientists, readily concede the possibility of minor adaptations to changing environments along the lines Darwin outlined. But some object that the spawning of fundamentally new biological forms is necessarily a teleological project in pursuit of a new physiological *goal* and therefore depends on the prior *conception* of a mind. To suppose otherwise, they contend, is a supposition requiring not just "imagination" but a truly heroic suspension of disbelief. This is because Darwin's idea essentially proposes that purpose can be achieved purposelessly.

Here we would do well to recall that rather jagged fault line between micromutations (the flounder acquires different colorings) and macromutations (e.g., the flounder slowly evolves into a more complex fish possessing new body parts with fresh capacities). That fault line has been excavated by critics and defenders alike in the wake of the *Origin*, and divides scientific opinion to this day. Some experts remain unpersuaded that all animal life could have evolved by the supremely chancy ministrations of Mother Nature alone, since—according to the terms of neo-Darwinian theory—natural selection must necessarily wait upon genetic mutations before it can kick in and begin its winnowing operations. The large role accorded to chance in the evolutionary process has often been viewed as standing in considerable tension with the notion of a predictable *mechanism*, a consideration which impeded acceptance of Darwinian ideas.

Chance in Darwin's Theory

For the first decades of Queen Victoria's reign, any would-be scientific law dependent on the postulation of chance would have condemned itself as being a contradiction in terms. For chance, by definition, is not a causal *agency* nor does it adhere to the predictability and regularity of a natural law.

It's true that Darwin likely conceived of the chance element of his theory not as randomness at an ontological level, as for instance randomness conceived of in some formulations of quantum mechanics, but instead as random only in an epistemological sense, as referring to events whose causal chains are unknown and perhaps indeterminable, or at least not describable by current natural law, as well as not arising for the sake of the survival and reproductive needs of the organism. As Darwin explained in the *Origin*, "I have hitherto sometimes spoken as if the variations—so common and multiform in organic beings under domestication, and in a lesser degree in those in a state of nature—had been due to chance. This, of course, is a wholly incorrect expression, but it serves to acknowledge plainly our ignorance of the cause of each particular variation."[3]

But is the term wholly incorrect? Think of a rock dislodged by happenstance, tumbling down a mountain, and coming to rest in some particular spot. Yes, if one were all-knowing, one could provide a detailed, mathematical description of the natural forces accounting for the rock's journey—its every roll and skip, its particular rate of acceleration, its particular orientation when it came to a stop, and so on. But as this far outstrips our capacity to manage, we remain ignorant of its full causal story. And yet, even aware of that ignorance, we can hardly be said to abuse the language when we describe the particulars of any such haphazard journey down the mountain as random or chance-like, even recognizing that such natural forces as gravity, inertia, and friction played their parts in the rock's precise journey. In this regard, Darwin has done little more than state the obvious as regards what he means by chance. His attempt to distance himself from the conventional meaning of the term "chance" calls to mind the famous line from Shakespeare: "Methinks the lady doth protest too much." Darwin, it seems, was acutely aware that the prominent role of chance in his chance/selection theory was not well calculated to impress his Victorian colleagues.

Some modern, popular-level accounts of Darwin's ideas give the impression that with his theory of evolution a Great Discovery had been made whose self-evident brilliance was such that every educated person not blinded by religious commitments readily assented to it. This reading of Victorian history, however, is not supported by the evidence. The common opinion of the leading men of science in the first half of the nineteenth century tended towards a form of deism, according to which divine law initiated life on earth. More broadly, educated opinion had, by the 1830s, become comfortable with the notion of a remote God having acted indirectly through natural laws of His own design.

The result was that the predisposition in the natural sciences was in favor of natural laws and against the invocation of either miracle or chance. And so it was that Darwin's unheralded announcement of a process in which chance variations introduced all the creative steps leading to the whole panoply of terrestrial life struck many in Victorian England as counterintuitive at best. The idea that random variations (today understood as genetic mutations) lay at the root of a process that gave rise to the most intricate of designs was sharply opposed by eminent scientific figures such as Darwin's accomplished Cambridge tutor William Whewell and leading British scientist Sir John Herschel, the latter of whom reportedly described Darwin's chance-dependent theory as "the law of higgledy-pigglety."[4]

They were far from alone in their negative assessment. Prior to the *Origin's* publication, Whitwell Elwin, distinguished editor of the *Quarterly Review,* was selected by Darwin's publishing house to vet his manuscript in advance of possible publication. Elwin, not mincing words, memorably advised against publication on the grounds that he deemed the work a wild and foolish piece of imagination whose author would have been better advised to confine himself to the subject of pigeons.

Geologist, physiologist, and mathematician Samuel Haughton, a professor at the University of Dublin, sounded a similar warning, albeit in weightier tones. Haughton had become privy to the pre-publication preview of the *Origin* in July 1858 at a meeting of the Linnean Society convened in London by Charles Lyell and Joseph Hooker.[5] This was the occasion when, in the interests of gentlemanly fairness, the near-identical evolutionary views of Darwin and Wallace were given equal time.

In his autobiography Darwin makes special mention of Haughton, candidly reporting the Irish academic's verdict as having been that "all that was new in there was false, and what was true was old."[6] This was in fact a painfully honest representation of remarks Haughton really did make to members of the Geological Society of Dublin on February 9, 1859[7] and which he developed in an anonymous article written for the 1860 edition of *The Natural History Review*.[8] Haughton summed up his verdict on the papers by Darwin and Wallace (read out in absentia in both cases by Sir Charles Lyell) in unsparing terms: "There is no folly that human fancy can devise, when truth has ceased to be of primary importance, and right reason and sound logic have been discarded, that has not been produced, and preached as a new revelation."[9]

Haughton felt that Darwin had fallen for the logical *post hoc ergo propter hoc* fallacy, that is, as Haughton phrased it, the delusion that mere succession necessarily implied causation.[10] Zoologist St. George Mivart, while later to the show, was also to oppose the inference that similarities in physiological structure necessarily implied a phylogenetic relationship (a thesis Darwin would double down on in his second major work, *The Descent of Man*.[11]) In his 1871 book *On the Genesis of Species*, Mivart argued that such an inference involved circular reasoning: Similarities of bodily structure were simply *assumed* to be indications of common descent without concrete evidence. He felt that this was a circumstantial and superficial inference without clinching empirical support or even the prospect of confirmatory evidence emerging in the future.[12] Later in the same work Mivart dismissed Darwin's theory as depending on "mere minute, fortuitous variations."[13]

Other reviewers of the *Origin* also remained unconvinced by Darwin's claims.[14] The French Academy of Sciences was especially pointed, declining to elect Darwin to its zoological section on the grounds that his case for evolution was "not science, but a mass of assertions and absolutely gratuitous hypotheses, often evidently fallacious. This kind of publication and these theories are a bad example which a body that respects itself cannot encourage."[15]

Such, by and large, was the kind of professional response the *Origin* received. Mivart and Haughton, in company with members of the wider Victorian public,[16] persisted in regarding poor chance, with or without the aid of natural selection, as unequal to the task of bringing such a grand and exquisitely crafted symphonic whole as the terrestrial biosphere into existence without any agency to compose and direct it.

"A Deliberate Campaign"

The pre-publication slights and critiques Darwin received had the effect of putting him on the back foot. How could he get an idea so offensive to accepted scientific tenets embraced by both the scientific community and the wider public? What was his solution? Curtis Johnson took up this question through detailed researches into Darwin's private notebooks, letters, and the multiple editions of the *Origin*. In his 2015 book *Darwin's Dice,* Johnson concludes that Darwin increasingly de-emphasized the idea of chance as a significant causal factor in evolution, especially after the first edition of the *Origin*. By the sixth edition, in response to criticisms received, the theory had been reframed to emphasize the guiding hand of natural selection, making variation seem less random and more directed.[17]

Darwin's private writings reveal that for later, revised versions of the *Origin*, he massaged his material as part of a sustained effort to get more of his readership to swallow the suspect pill of chance. This massaging is most apparent in the sixth and final edition. By then, forewarned and forearmed by Mivart's verbal squibs about mere fortuity, Darwin had, according to Johnson, "adopted a variety of rhetorical strategies that added up to a deliberate campaign to retain chance as a central element while making it appear to most readers that he did not."[18]

In other words, Darwin became steadily convinced of the necessity to *insinuate* his dangerous idea into the consciousness of his peers by such means of verbal dexterity as he could devise, contriving to bypass his peers' critical guard in a campaign of *trompe-l'oeil* or, in plain English, of pulling the wool over people's eyes. Throughout his career, Darwin was never at a loss to construct imaginative and ingenious hypotheses. (See Chapter 3 below.) He would now put to hard service that aptitude as he strove to de-emphasize the role of chance in his theory.

By such an expedient Darwin hoped that criticism of his theory might quickly evaporate, a hope that his then-ally Alfred Russel Wallace appeared to share when he advised Darwin to delete the word "accident" and replace it with some such bland circumlocution as "variations of every kind are always occurring in every part of every species."[19] Such careful locutions were to become part and parcel of a studied policy of obfuscation.

Natural Selection: The Other Half of the Deal

One might suppose that whereas the chance-variation side of Darwin's joint mechanism of evolution was a tough sell, at least the ingenious other half of the mechanism, natural selection, posed a relatively easy marketing challenge. But natural selection faced its own public relations difficulty—empirical evidence for it accomplishing anything of note was in short supply.

Here too Darwin had recourse to a sleight of hand, using the metaphor of an architect to describe the work of natural selection. However, as Johnson has pointed out, the architect metaphor worked at cross-purposes with Darwin's intentions, since an architect is manifestly an intelligent agent, in contradistinction to the witless and impersonal processes of natural selection.[20] To make matters worse, the image had been employed for centuries in such locutions as "cosmic architect" to refer to the very deity Darwin wished to exclude.[21]

Clearly the take-away message from his writing, however deftly and euphemistically camouflaged, *does* involve the conception of the chance origin and evolution of life on this planet. On that reading

God had been shown the door, having been rendered superfluous to evolutionary proceedings said to unfold autonomously. Yet Curtis Johnson also observes that in notes *not* for public consumption Darwin would ask himself, "Do these views make me an atheist?," whereupon he exclaimed "NO!" In later notes he describes himself variously in theistic, deistic, or agnostic terms. Such terms, Johnson notes, "preserve the possibility... of a Creator who designed a world in the beginning that would operate in definite and predictable ways."[22]

At a later point in his life, Darwin may have been seriously tempted to return to the Christian fold, at least with one foot, as will be observed below. Perhaps even more tempting would have been a decision to return to the status quo ante, that is, to the common deistic prepossessions of the scientific community in the first half of the nineteenth century. Such a reversion might have been the easier for Darwin since, in the case of men of science, their convictions rested not on simple faith but on two centuries of philosophical thought. Those new patterns of thought had brought such enormous changes to people's understandings of external nature as to make Darwin's twin mechanisms of chance and natural selection problematical, if not indefensible. It is to that fundamental change in the educated European mindset that we now turn.

Changing Ideas of Nature

In the two centuries which separated William Shakespeare from Charles Darwin, there had occurred a revolution in Europe's understanding of nature—a change such as to make improbable the very notion of natural selection. It is well enough known that, in the early decades of the seventeenth century, Shakespeare could quite unproblematically portray an active nature capable of issuing portents of future misfortune as warnings to humanity. On the stage of the Globe theatre, it will be recalled, nature was able to put forth such meteorological warning signs as thunderstorms and sundry other "horrid sights seen by the watch." But such stage evocations are interpreted by today's audiences as merely a dramatic device (known technically as the "pathetic fallacy"). For hard on Shakespeare's heels chronologically came a finger-wagging brigade of mid- to late-seventeenth-century natural philosophers exhorting the bard's intellectual heirs to view nature as an entirely passive phenomenon. No will or agency should henceforth be attributed to natural phenomena, these early scientists insisted, and this understanding became an informing axiom of post-Enlightenment science.

As Stanford history professor Jessica Riskin has pointed out in *The Restless Clock: A History of the Centuries-Old Argument over What Makes Us Tick*,[23] a residue of the older thinking could still be traced in Lamarckian conceptions that posited an immanent force driving plants and animals to form themselves and to "complexify" their structures over time to produce many different plant and animal types. Harking back to ideas originally proposed by Denis Diderot and like-minded *philosophes* in mid-eighteenth-century France, Lamarck proposed that living beings might be able to alter their own internal physiology and, by an act of will, develop such new organs as might be requisite to their purposes in life. His most (in)famous illustration of this claimed ability was the much-cited (and derided) example of the giraffe species possessing the ability to elongate its neck over the course of many generations to be able to reach leaves at the tops of trees.

Consistent with the tenets of this idea of a power inhering in nature itself, Lamarck was to look askance at the famous watch-on-the-heath analogy formulated by William Paley, wherein Paley compared the functional arrangement of parts in a watch to the same in a living organism and argued that just as the one is necessarily designed, so is the other.[24] Lamarck demurred, positing that the essential motor of the living being was, unlike with a watch, integral to it rather than something coming from an outside source. More than this, Lamarck was suggesting, inhering in living things is some natural force affording a species the capacity to evolve, given the right circumstances.

Lamarck's was then a decidedly bottom-up conception and, as Riskin observes, Lamarck's was a "dangerous" idea since it posited *self-evolving* entities. Lamarck appeared to have replaced an external God with a mysterious and unspecifiable inner force. A common contemporary take is that in rejecting Lamarckism, science rejected what amounted to an attempt to smuggle back into biology an antiquated animism. Riskin offers a decidedly different perspective. On her view, "Lamarckism represented its era's most naturalist, nontheological account of species-change."[25]

This means, explains Riskin, that the biologists and philosophers who dismissed Lamarck's theories were unwittingly reinforcing a view most of them had no desire to support. They were placing themselves in the strangely anomalous position of distancing themselves from naturalism and, in so doing, becoming heirs to the very tradition with which they meant to do battle: the argument from design, with roots among the ancients and revived and further developed in the physio-theology of the seventeenth, eighteenth, and early nineteenth century, championed by such thinkers as John Ray, William Derham, Christian Wolff, Bernard Nieuwentyt, and latterly, William Paley. As Riskin puts it, "A material world lacking agency assumed, indeed required, a supernatural god."[26]

The seventeenth-century banishment of agency, perception, consciousness, and will from nature gave a monopoly on all those attributes to a supreme, external power: The paradigm simply could not function without "an accompanying theology."[27] It was an unambiguously top-down conception which depended absolutely on the superintending agency of a designer-creator.

This, then, is the grand irony: Darwin and his heirs had hitched their wagons to a tradition that from its inception denied agency to inanimate nature *precisely in order to ascribe it to a designer God*. The framework to which Darwin expressed his official allegiance was, then, *ab initio* and by the express wish of its originators, extremely design-friendly.

By restricting mysterious agencies to the jurisdiction of the Christian God, the shapers of modern science (after circa 1650) left the atheistically inclined with a dilemma. The new, mechanistic science was adamant that one could not procure a lens instrument without an instrument maker, or an eye without a divine Optician (hence Darwin's famous shudder when he asked himself rather disconsolately how anything of such supreme intricacy as the eye could have been fashioned by natural selection).[28]

In short, Darwin found himself in a bind. He sought a mechanism for evolution without recourse to either of two resources available to evolutionary thinkers of the preceding two centuries. If we can allow ourselves a bit of picturesque language in summarizing the situation, Darwin sought to bar nature's Father from evolution's work, but in seeking a power that was up to the creative task, he could not, as his grandfather or Lamarck had done, go running back to Mother Nature.

Of Algorithms and Waving Statues

As we have seen, Darwin's solution was to appeal to chance plus natural selection, and then to (a) obscure as much as he could the central role of chance in the operation, and (b) smuggle in a hint of planning and purpose by speaking of nature "selecting" and of natural selection as an architect.

Darwin, to his credit, at least had the grace to harbor doubts about the role of chance in evolution, which is more than can be said of many of his twentieth- and twenty-first century legatees. Here is Daniel Dennett expounding with unruffled finality (and circularity) on what he terms his "algorithmic" ideas about natural selection: Can the biosphere "really be the outcome of nothing but a cascade of algorithmic processes feeding on chance? And if so, who designed that cascade? Nobody. It is itself the product of a blind, algorithmic process."[29]

An even more remarkable computation of the power of mere chance can be found in Richard Dawkins's *The Blind Watchmaker* where, in the context of assessing whether certain phenomena might be adjudged impossible or merely improbable, Dawkins seriously moots the (albeit remote) possibility of a marble statue moving its arm:

> In the case of the marble statue, molecules in solid marble are continuously jostling against one another in random directions. The jostlings of the different molecules cancel one another out, so the whole hand of the statue stays still. But if, by sheer coincidence, all the molecules just happened to move in the same direction at

the same moment, the hand would move. If they then all reversed direction at the same moment the hand would move back. In this way it is possible for a marble statue to wave at us. It could happen.[30]

I confess that on first reading that paragraph I did not know whether to laugh out loud or question my own sanity ("Did he *really* just write that?!"). The latter worry was in fact only fully allayed when I came across the volume entitled *Answering the New Atheism*, where I discovered with some relief that I was not alone in finding the notion absurd. The two authors of the above volume, Scott Hahn and Benjamin Wiker, quote the same paragraph because, "if we merely reported it, no sane person would believe that Dawkins had written it." They continue:

> Our concern for now is whether Dawkins' unconquerable faith in the powers of chance is rational. For Dawkins, whatever God could do, chance could do better, and that means that any event, no matter how seemingly miraculous, can be explained as good luck....
>
> And if such impossible things are possible, why isn't it possible that it was indeed a miraculous occurrence? Why isn't the miraculous itself a possibility?[31]

Quite so. Wiker and Hahn are to be commended for pointing out what most modern reviewers of Dawkins's volume failed to flag up. Assuming that the critics' silence did not simply spring from obscurantism, one can only speculate that their collective genuflection was motivated partly by materialist confirmation bias and also conceivably by a form of intellectual doffing of the cap to an Oxford grandee. Effectively it is as if the reviewers had been caught in the headlights of a car, which froze their critical faculties and rendered them incapable of pointing up the absurdity of Dawkins's arm-waving statue.

Fortuna Miscast

It can at least be said in Dawkins's defense that he was building on a long tradition. Luck or chance in her personified form as the goddess Fortuna revolving her wheel (the infamous *rota Fortunae*) was often portrayed in the iconography of the medieval world as the ficklest of the divinities of the classical pantheon, matched only by the two-faced Roman god, Janus.

Not for nothing did Geoffrey Chaucer write in his fourteenth-century "Knight's Tale" of "Fortune and hire false wheel,/That noon estat assureth to be weel" [Lady Luck and her untrustworthy wheel which guarantee *nobody's* good fortune]—a quotation which has stayed with me since school days, since it expresses a bitter truth we have surely all been obliged to taste on many occasions. It is grimly appropriate that Fortuna's wheel became the prototype of that ruin of so many, the modern roulette wheel. But even though chance was apostrophized in the medieval *Carmina Burana* as *Fortuna Imperatrix Mundi* (Fortune, the world's empress) she was never described as *Fortuna Creatrix Mundi* (Fortuna, the creator of the world). To be asked to believe that the biological equivalent of the invariably untrustworthy Lady Luck was in good part responsible for the evolution of all organic life is a big ask and, I fancy, an idea which our medieval predecessors would have laughed out of court.

There are many further objections to Darwinian theory which, it should be noted in passing, have a strictly scientific/logical basis and have nothing to do with either cultural zeitgeist or religion (contrary to what has been persistently argued by atheist biologist Richard Dawkins).[32] In fact most objections rest on evolutionary theory's lack of evidential foundation and explanatory force—a lack becoming less deniable with every passing decade. Ever since the discovery of DNA in 1953, we have known that the "simple life" originally invoked by Darwin and his German follower, Ernst Haeckel, as the base of the evolutionary tree is in reality a phenomenon of such limitless complexity as to lie, even to this day, beyond our comprehensive understanding. No less a person than Bill Gates, the founder of Microsoft, has declared that "DNA is like a computer program but far, far more advanced than any software ever created."[33] And DNA is but a small part of the many-layered complexity of even the simplest single-celled organism.

With the cracking of the genetic code, the problem of origins has shifted largely to the mystery of the origin of biological information.

For life is now known to be all about information and information-processing systems, both genetic and epigenetic, and all essential to the unfathomed high technology of the molecular biological realm. As Eric H. Anderson puts it: "Not just complex. Complex and co-ordinated, with a… digital code, information storage, retrieval and translation mechanisms, error-correction algorithms, functional integrated systems, and molecular machines—marvels of nanotechnology that put to shame anything humans have yet created."[34]

And as with complex, functional systems of human technology, the many essential parts tend to be highly interdependent and tailored one to another, in such a way that replacing them one small change at a time becomes impossible while maintaining function. For a machine being redesigned in a human factory, this poses no great difficulty. The machine being redesigned can survive such periods of dysfunction, for it was never alive in the first place. But any organism with any prayer of evolving some new technology (e.g., the capacity for winged flight) carries the added burden of very much needing to remain alive in order to participate in the evolutionary chain of progress, at least long enough to mature and breed.

Hence, an increasing number of scientists today have little faith in the existence of as yet unspecified evolutionary pathways towards the biological goals invoked by ardent evolutionists. If anything, such vaguely imagined pathways now seem less navigable, and thus less plausible, than they did a half century ago, when so much of the complexity noted above had yet to be uncovered.

This increasingly acute challenge notwithstanding, the official story propagated in high school biology textbooks and in the popular media is that Darwinism was melded unproblematically with Mendelian genetics in the early 1940s, and all remains right with the world of Darwin. The updated form of Darwinism no longer speaks vaguely of random variations. Instead, the neo-Darwinian speaks of "genetic mutations" that enhance survival and reproductive capacity within a particular environment. The individual with such a mutation would tend to outbreed other members of its population, leading to those mutant genes proliferating in the population. Rinse and repeat through numerous genetic mutations and numerous generations until an entirely new biological form has evolved.

In reality, this neo-Darwinian story faces an acute problem. There appear to be innumerable cases where multiple chance mutations are needed in one go to produce a functional breakthrough, and just the right sort of genetic mutations. Lehigh University biologist Michael Behe detailed several such instances in *Darwin's Black Box* (e.g., the blood-clotting cascade, the bacterial flagellum, vision) and ably defended his argument against attempts to explain these cases away.[35] Since then many other such entities have been added to the list.[36] Perhaps more damningly, there remains a standing invitation from Behe and other design theorists for Darwinians to produce even a single detailed, causally credible description of an evolutionary pathway for gradually evolving any complex biological machine or system. The invitation, as of this writing, appears to remain unanswered.

That's the challenge from molecular biology. The neo-Darwinian story also faces a challenge from paleontology, one that confronted Darwin's original theory. Darwinism, both old and new, anticipates biological forms very gradually evolving into new forms and thus leaving behind numerous transitional fossils in the geological column. But such transitionals are sorely lacking.

Darwin blamed the incompleteness of the fossil record for its failure to confirm his theory, and anticipated that caches of confirmatory fossil evidence would be disinterred after his death. That hope has not been realized. If anything, the challenge of the fossil record is more acute now than in Darwin's day.

One may have heard the phrase "the missing link." The term is misleading. The fossil record contains not one or even a mere handful of "missing links." Rather it contains, in the words of Geoffrey Simmons's cheekily titled book, *Billions of Missing Links*. Up and down the geological column, the transitional forms that both Darwinism and neo-Darwinism anticipate are absent.

Darwin's appeal to an incomplete fossil record did not persuade even his greatest ally, Thomas Huxley. Despite his indefatigable partisanship for Darwin, Huxley would assent only to the phenomenon of evolution, but *not* to Darwin's distinctive innovation of natural selection working on minute variations and an unwavering gradualism. What worried Huxley was the dearth of fossil remains to chart the claimed slow evolution and gradual speciation of the biosphere. "The combined investigations of another twenty years," stated Huxley, contemplating the theoretical possibility of more convincing fossils eventually being disinterred, "may, perhaps, enable naturalists to say whether the modifying causes and the selective power, which Mr. Darwin has satisfactorily shown to exist in Nature, are competent to produce all the effects he ascribes to them; or whether, on the other hand, he has been led to over-estimate the value of the principle of Natural Selection."[37] So Huxley would not, he repeated in a speech in 1880, accept the theory of natural selection until further paleontological work had made the case incontrovertible.

As Huxley saw matters, the doctrine of natural selection presupposed evolution, but evolution need not entail natural selection, or for that matter, gradualism. How he expected evolution to make large and rapid leaps, or to filter useful from deleterious variations in order to achieve forward progress, all without the aid of natural selection, he never made clear. Huxley's reservations, however, proved well founded. Although the fossil record does document a general trend from the less complex billions of years ago to the more complex, the leaps in complexity do not show up gradually, as the Darwinian model urges.

This was the testimony of the fossil record in Darwin's day, and it remains the testimony of the fossil record into the twenty-first century. Indeed, after more than a century of additional fossil finds, the testimony is all the more insistent on this point. Biologist and NIH Distinguished Investigator Eugene Koonin thus observes: "The relationships between major groups within an emergent new class of biological entities are hard to decipher and do not seem to fit the tree pattern that, following Darwin's original proposal, remains the dominant description of biological evolution."[38]

A Theory Split Down the Middle

In 2007, in the same article quoted immediately above, Koonin memorably observed, "Major transitions in biological evolution show the same pattern of sudden emergence of diverse forms at a new level of complexity."[39] This comment did not come out of the blue. Thirty-five years before, in 1972, it was already clear to many that the abundant fossils hoped for by Darwin probably did not exist. American paleontologists Stephen Jay Gould and Niles Eldredge, hoping to loosen the then-century-old biological Gordian knot, put forward an alternative to the Darwinian narrative of slow evolution, one involving development by leaps and bounds. They termed their modified Darwinism "punctuated equilibrium."

On this view, there was a greater likelihood of radical changes leading to new species in places where animal groups had become isolated geographically. In the more favorable cases, Gould and Eldredge theorized, dramatic spurts, or what historically had been called saltations, occurred, followed by protracted periods of stasis, in which the evolutionary process was largely dormant. Only on this understanding, they argued, could the numerous punctuations (gaps) in the fossil record, along with the fossil record's abundant testimony of long periods of stasis (equilibrium) be made sense of. Gould would say that natural selection was still governing the evolutionary process, but his formulation represented a stark break from Darwin's vision of gradual evolution recorded in a seamless procession of fossil intermediates throughout the history of life.

As we have seen, Darwin acknowledged and indeed drew attention to the lack of fossil evidence for his theory, and remained puzzled till the end of his life as to how certain of nature's more complicated structures could have evolved via his proposed mechanism. But it appears that he was so psychologically committed to his own theory that he was unable to follow such evidence where it seemed to point. That evidence, looked at dispassionately, suggests that most new kinds of organisms have appeared unheralded and without forerunners, arriving on the scene in a way directly opposed to that postulated by Darwin.

Take the paradigmatic example. Some 540 million years ago there occurred a phenomenon known as the Cambrian explosion, involving the sudden emergence on Earth of about thirty distinct phyla. Phyla represent the highest taxonomic level of animals, each phylum representing a general body plan radically distinct from those of other phyla. To give some idea of how broad a class a single phylum is, humans, bears, rabbits, fish, frogs, crocodiles, and birds, despite their substantial differences, are all part of not only a single phylum but a single subphylum. That tells you how significant the differences must be between different phyla. And again, thirty distinct phyla appeared relatively suddenly in the Cambrian explosion. No transitional intermediates were found from this era, and there is no indication that this menagerie of animal forms did *not* emerge fully-formed. Indeed, the phenomenon has all the appearance of a biological Big Bang.

The Cambrian explosion, moreover, is merely the most dramatic of many such biological Big Bangs in the fossil record.

It is this fossil record pattern, anomalous within a Darwinian explanatory context, that the Gould/Eldredge theory of punctuated equilibrium was intended to explain. Courteously (or perhaps prudently), neither scientist challenged the basic principle of Darwin's theory. That is, they did not propose a new basic mechanism for evolution. The innovation, rather, was an emphasis on the role of isolated populations to explain why the Darwinian mechanism seems always to do its most dramatic work in a manner invisible to the fossil record.

Gould and Eldredge's genuflection to Darwinian orthodoxy notwithstanding, it is clear that their theory, if it is to provide any relief from an uncooperative fossil record, requires evolution to work far more rapidly (and from within far smaller populations) than a thoroughly orthodox neo-Darwinism ever anticipated, and thus did indeed mark a radical attack on the theory's cardinal principle of gradualism.

For the last half century there have then been two principal variants of modern Darwinism in circulation, the gradualistic version of the mainstream, and the Gouldian variant optimized to make sense of the fossil record, and nobody has yet succeeded in definitively adjudicating between the two. Marcel-Paul Schützenberger dismissed both plain vanilla neo-Darwinian and the Gouldian variant as examples of "highly competent people" discovering "a number of poetic but essentially hollow forms of expression."[40] For him both theories are human confabulations which do not correspond to anything "out there" in the world of empirical reality any more than the geographical distortions of the Hereford Cathedral *Mappa Mundi* could be classed as a cartographically precise illustration of the medieval world.

The fundamental problem with the theory of punctuated equilibrium remains precisely the one identified long ago by Darwin himself when he argued that nothing in nature could manage rapid bursts of creativity in the history of evolution. No, the Gouldian variant does not envision instantaneous bursts of significant creativity, but the Gouldian variant, unlike the Darwinian version, cannot even appeal to vast swaths of time as its co-adjutant, nor to impressively numerous breeding populations. And although deep time as a factor in evolution has been criticized for being something of a cop-out and misleading canard,[41] it was, for good or ill, integral to Darwin's thinking.

Darwin insisted on an almost imperceptible rate of evolutionary growth since he was aware, long before the modern discovery of molecular biological complexity, that the co-adapted construction of organisms constrains the extent to which one organ can change without correlated changes in other organs of the body. After all, animals are not constructed on the modular principle of simply fitting body parts together in the manner of prefabricated low-cost housing or Meccano sets. Imagine, as Darwin did, going from a hypothetical short-necked proto-giraffe to its long-necked descendant. To pull off such a transition, evolution would need to co-modify the animal form's spinal discs, its postural reflexes, and its heart and circulatory system, not to mention its evasive tactics vis-à-vis predators as its neck lengthened. The elongating, if it were to provide any biological advantage to the animal, must needs be only one part of a *coordinated* process of modification in order to be physiologically sustainable. Such detailed organizational choreography would, if possible at all, require eons —and the Gouldian variant does not allow for such a protracted process.

So if Darwin were to have assented to the postulate of punctuated equilibria, he would have been obliged, following his own logic, to accept something like the account of creation found in the Bible or in the seventh book of Milton's *Paradise Lost*, which has God commanding the earth to bring forth created beings *ex nihilo*.[42] Such a concession would of course have thrown Darwin back into what he saw as the metaphysical morass he'd been trying to escape in the first place; hence his insistence on a positively "Jurassic" pace of evolution *sans* saltations.

If we imagine Darwin living to the 1970s, it isn't difficult to also imagine him concluding that Gould and Eldredge were pushing a crypto-biblical theory sharply at odds with his own wholly naturalistic postulates. Nevertheless, however one may view this bifurcation of Darwinian theory, the notion that there is only one normative version of neo-Darwinism, with evolutionists only differing over the details, is untenable. Such a fundamental disagreement in the ranks of evolutionists lends support to Michael Denton's diagnosis some four decades ago of a crisis in evolutionary studies. The dispute initiated by the arrival of Gould and Eldredge's punctuated equilibrium model, moreover, was but the harbinger of other fundamental disputes among evolutionary biologists, disputes generally kept out of the public eye but spilling into the public square with the convening of a group of mainstream evolutionists at the venerable Royal Society of London in 2016 to explore, and frankly confess the need for, a suitable alternative to neo-Darwinism. The search for a "third-way" or "extended evolutionary synthesis" that delineates an adequate cause for the origin of life's great diversity remains ongoing, even as the various contenders ably point up the shortcomings of each rival materialistic theory.[43]

Quite aside from scholarly disputes, however, there was one problem Darwin found so daunting that he elected not to confront it at all in the *Origin*, and only glancingly in his private correspondence. That issue concerned the question of the *absolute* origin of life on earth.

First Life

Arguably the greatest problem for the acceptance of modern evolutionary theory as a self-standing and logically complete account is the unsolved mystery of life's ultimate origin, a subject which Darwin tried to bat away as, if not a total irrelevance, then as something beyond his competence to pronounce on. Even today Darwinian supporters will downplay the subject of the absolute origin of life as a matter extraneous to the subject of modern Darwinism. It was not the focus of the *Origin* or any of Darwin's later books touching on evolutionary theory, but the question of the origin of life is absolutely foundational to the integrity of any evolutionary framework purporting to offer a comprehensively naturalistic account of the origin of Earth's biosphere.

Nineteenth-century German scientist Ludwig Büchner called Darwin out on this. If it could be admitted that a creator had "breathed into" being the first living organism or organisms, as Darwin, referencing the ancient doctrine of the divine *pneuma*, worded it,[44] why should not special acts of that creator be admissible for various subsequent stages of human and animal development?[45] (At least on that allowance the indubitably teleological nature of the selective process would find a logical foundation that up to that point it lacked.)

The response among those naturalists who took Büchner's point to heart was to search for some means by which purely natural processes might indeed generate the first life. The search eventually moved into the lab, but it began more modestly and with Darwin himself. For naturalists of the Victorian period, the problem had only recently become acute, with the demise of the once widely held notion of spontaneous generation. Not long after the *Origin* first appeared in print, a laboratory experiment by Louis Pasteur confirmed early work by Sir William Harvey and Italian scientist Francesco Redi, putting to rest once and for all the idea that life routinely and spontaneously sprang from non-life. The question of life's origin was now undeniably an unsolved problem.

Darwin tackled the question in a letter to his friend Joseph Hooker, speculating that the first life might have emerged spontaneously in a "warm little pond," one just happening to have just the right chemical ingredients, and the nascent life helped along by the fact that there were no other organisms yet around to attack and devour it.[46] His just-so story was the vaguest fog piled on fog and ignored the clear trend toward a growing appreciation for how difficult, and perhaps impossible, it was for nature on its own to make the great leap from a dead planet to one with life.

Abiogenesis to the Rescue

Others would follow in Darwin's footsteps on this point. In what was shaping up to become the largely post-Christian twentieth century in Europe, it was hardly unexpected that the death of spontaneous generation should have been stubbornly denied on ideological grounds. The secularizing trends of the early twentieth century meant that the dead idea of spontaneous generation was kept around, zombie-like. Somewhere along the way the idea was rechristened "abiogenesis," presumably to escape the tincture of obsolescence and decay that hovered over the older term "spontaneous generation."

Russian biologist Alexander Oparin and British scientist J. B. S. Haldane led the way with speculative work on the subject published in 1924 and 1929 respectively. Fruitful laboratory work had to wait until 1953 and the now famous Miller-Urey experiment at the University of Chicago in 1953.

The unspoken hope behind this now (in)famous experiment was the possibility that Pasteur, Harvey, and Redi might have been wrong and that mid-twentieth-century advances might discover a solution where predecessors had failed. To be fair to Stanley Miller and Harold Urey, this was no desperate shot in the dark. Miller and Urey (the latter a later recipient of the Nobel Prize) theorized that if the conditions prevailing on primeval earth were reproduced in laboratory conditions, such conditions might prove conducive to a chemical synthesis of living material. Accordingly, an electric spark was passed through a mixture of methane, hydrogen, ammonia, and water to simulate the kind of energy which might have come from thunderstorms on the ancient Earth. The resulting liquid turned out to contain amino acids which, though not living molecules themselves, are the building blocks of proteins, essential to the construction of life.[47]

After these results were announced, hopes ran high that complete experimental support for the idea of abiogenesis was just around the corner. Surely researchers were on the cusp of extending Darwin's evolutionary story back to the prebiotic Earth and so vindicating Darwin in his effort to provide an account of the origin and diversification of life on earth that was both causally adequate and free of the intervening activity of a divine creator. Such hopes, however, have not been realized.

Today, more than seventy years later, the goal appears, if anything, much further away than it did in 1953. The first sign of trouble arrived in the very same year that the Miller-Urey experiment took place, when Francis Crick and James Watson, building on experimental results by Rosalind Franklin, succeeded in identifying the famous double helical shape of DNA. Their discovery, and those that quickly followed in its wake, made clear that getting life from non-life wasn't principally about conjuring up the various requisite chemicals, even if the presence of such chemicals was certainly a necessary condition for life. Watson and Crick's breakthroughs made painfully clear that life required not just the chemicals but code. Just as one might have the precise ingredients to make a book (ink, paper, glue, etc.) and still not have a book, so too one might have all the requisite chemical ingredients for life and not have life. A crucial missing ingredient in both cases: information.

Darwinism's Persistent Problems—Legion

This foray into the question of the origin of life should not suggest to the reader that, having scratched the bottom of the barrel, I was forced to turn to a problem outside the theory proper in order to cast additional shade on the gentleman from Down House. Not at all. There are numerous further problems that might be discussed with regard to what even Darwin himself in the sixth chapter of the *Origin* referred to as the "difficulties" facing his theory. For instance, the idea of a crossover from one species to another appears particularly problematical in view of the practical experience of animal husbandry. Darwin appealed to artificial breeding to support his theory, but the original objections to this move have never been satisfactorily answered, namely that domestic breeding is a case of artificial rather than natural selection; and even with the benefit of intelligent guidance, it has never succeeded in fashioning fundamentally new biological forms from existing ones. Dogs remain dogs, cats remain cats. Moreover, this empirical result is now well understood at a conceptual level: The genetic code possesses what might for convenience be termed an inbuilt fail-safe system to ensure genetic homeostasis and the integrity of individual species, a fail-safe likely indispensable due to deleterious and lethal valleys in what are known as fitness landscapes, valleys on any theoretical evolutionary pathway from one biological form to a distinctly different one. I shall return to this and other objections where appropriate below.

The litany of objections is a long one, and I shall not attempt a systematic enumeration. The task has in any case been painstakingly carried out in many other publications, some of them laid out in matter-of-fact tones, others bristling with righteous indignation at what they portray as the Darwinian confidence game. Such works have proliferated to such an extent in the last few decades as to have formed something of a literary sub-genre. Among those responsible for their production are even mainstream biologists committed to the project of discovering a thoroughly naturalistic account of the origin and diversification of life on earth.[48] In fact, it would sometimes appear from the sheer numbers of persons rushing into print to point out the fallacies and evidential failures of Darwinism that the only class of persons purporting to find Darwin credible any more is a coterie of professional biologists with a lifetime invested in the Darwinian paradigm, together with lay persons and superficially trained biology instructors prone to an unreflective science-olatry.

How did we get to this point? And why do so many whose scientific expertise lends them immense cultural influence elect to use that influence to persuade people of a theory which struggles to withstand even a modest degree of rational critique? I endeavor to answer these questions in the chapters that follow.

2. Metaphors, Analogies, Subtexts

*T*HE ORIGIN OF SPECIES APPEARS IN ITS INCEPTION TO HAVE BEEN at least in part an interrogative discussion inviting a dialectic response rather than the materialist monolith it was taken to be after the "new synthesis" cemented its position with the imprimatur of hard-faced modernity in the early 1940s. In the nineteenth and early twentieth century Darwinism was not yet widely employed as a rallying cry to advance the cause of militant atheism. Some early readers even understood it as a theory which buttressed their Christian faith. Darwin himself invited such a response at the conclusion of the first edition, and still more pointedly beginning with the conclusion of the second edition. There he wrote:

> Thus, from the war of nature, from famine and death, the most exalted object which we are capable of conceiving, namely, the production of the higher animals, directly follows. There is grandeur in this view of life, with its several powers, having been originally breathed by the Creator into a few forms or into one; and that, whilst this planet has gone cycling on according to the fixed law of gravity, from so simple a beginning endless forms most beautiful and most wonderful have been, and are being, evolved.[1]

The passage has been widely celebrated for its eloquence. At the same time its theological overtones also have been long regarded as a sop by the author to religious readers of the book, a cherry coating on the bitter materialistic pill that was his naturalistic theory of biological origins. Be that as it may, there was a diversity of readings of the theory even among his inner circle of scientists and friends. Some found in the *Origin* room for what today has become known as theistic evolution, a reading that found favor with many in both Britain and America in the later Victorian period and beyond. Their perspective involved a tacit grafting on to Darwin's evolutionary process a thin but crucial layer of theistic superintendence, as James Moore documented in his standard study of post-Darwinian controversies.[2] And here the term "theism" is being used with care, since for many of these Victorian theistic evolutionists, the deity overseeing the evolutionary process was regarded as more hands-on than was allowed for in the minimalist conceptions of deism. For them, Darwin's random variations and natural selection played their parts, but God remained the ultimate choreographer. The author Charles Kingsley and the future archbishop Frederick Temple, for example, saw in Darwinism a form of progressive revelation: science illuminating for humanity what had once been hidden. Kingsley came to view biological evolution as a branch of what German theologians call *Heilsgeschichte*, that is, salvation history, according to which God constantly works behind the scenes to promote the human potentialities and ultimate salvation of His subjects. For Kingsley this hidden-hand approach seemed infinitely more satisfactory spiritually than the deist position which postulated a God who had made a once-and-for-all effort of creation but had since that time supposedly retired from his exertions with little more care for those he had created.

So downstream from the *Origin* we have atheistic Darwinists, deistic Darwinists, and theistic Darwinists, each claiming the theory for their own. Such diverse interpretations of the significance of Darwin's work prompt us to reconsider the definitional status of its key term, "natural selection." What precisely did Darwin mean by the term, and is the term conceptually coherent? Is natural selection a true process of nature or simply the reification of a thought experiment deserving of only notional status? In other words, does it, or does it not, make sense and so help humankind towards a reliable understanding of our existential situation?

What's In a Name?

No few of the problems which readers have registered on reading the *Origin* have been linguistic, arising from a curious lack of conceptual and definitional precision in Darwin's formulations. Darwin's somewhat confusing terminological choices are a feature of his work that even experts have had trouble with. In the English language prior to Darwin, "natural selection" was a term used by those who reared animals to denote nature's serendipity, those unpredictable changes to animal physiology over the generations arising from factors operating outside the sphere of the breeders' understanding and control. The term stood in semantic contradistinction to breeders' efforts to mate carefully chosen animals of each sex in order to encourage the emergence of favored features in the resulting progeny, what is often referred to today as "artificial selection."

Hence its *original* meaning in the common understanding of standard English was semantically anchored in notions of the random, unforeseeable and unplanned, something beyond the ken of breeders. However, when Darwin commandeered the term, a significant degree of semantic creep took place. Alfred Russel Wallace and other colleagues pointed out to him that there was little comparison between what animal breeders did by the use of human ingenuity and how mindless nature acted. Darwin conceded various differences between artificial and natural breeding, with some redounding to the creative powers of artificial selection, such as artificial selection benefiting from the foresight and goal-directedness of the human breeders. But he also insisted on advantages of natural selection over artificial selection. These included its greater timescales, its working on much larger populations, and the fact that natural selection worked toward the fitness benefit of the organisms themselves, rather than for those of the human breeders. In any case, he was unwilling to let go of his analogy to domestic breeders. In this anthropomorphic conception, Nature, like the human livestock breeder, exercised her own forms of selection. Darwin persisted in contending that Nature, with limitless millennia at her disposal, could do a more comprehensive job of bringing about major physiological changes (and eventually new species) than could intelligent human breeders. This is how he phrased that idea in the *Origin*:

> It may metaphorically be said that natural selection is daily and hourly scrutinising, throughout the world, every variation, even the slightest; rejecting that which is bad, preserving and adding up all that is good; silently and insensibly working, whenever and wherever opportunity offers, at the improvement of each organic being in relation to its organic and inorganic conditions of life. We see nothing of these slow changes in progress, until the hand of time has marked the long lapse of ages, and then so imperfect is our view into long past geological ages, that we only see that the forms of life are now different from what they formerly were.[3]

It is noteworthy in the above that the word "metaphorically" should have been used because it acknowledges that the process being described should not be taken literally but rather understood as a poetic image functioning as a conceptual aid to express something that could not otherwise be easily explained. Also noteworthy, the word "metaphorically" is not in the first edition of the *Origin* (1859), but was subsequently added by Darwin, presumably to protect himself (albeit not entirely successfully) against a charge of animism and the objection that he was supporting an anthropomorphic or even theistic understanding of the evolutionary process. Not without reason was Darwin's metaphor of natural selection recently decoded as "an anthropomorphic but superhuman agency, 'daily and hourly scrutinizing' all variation, and making intelligent and benevolent decisions like a Paleyan Designer."[4] The same point was made more than a century earlier by one of the leading lights of the Oxford Movement of the 1840s, Edward Bouverie Pusey, as the chronicler of English thought, Basil Willey, noted:

> This same metaphysical unawareness led him [Darwin] also, almost without noticing it, to replace the absent God with a latent personification of Nature, or even of "Natural Selection" itself. True, he catches himself out from time to time, and warns us that he is only speaking metaphorically when he talks of Natural Selection

"observing minutely," "with unerring tact discovering each improvement for further perfecting," and so forth. But he returns so habitually to that way of speaking, that we feel Pusey to be right in accusing Darwin of having himself introduced, into the theological vacuum he had created, a power acting according to design.[5]

The use of metaphors should not be dismissed as mere literary ornamentation. In this instance, Darwin's scientific discourse strays from providing a strictly materialistic account of speciation, with the language opening up—perhaps even encouraging—a non-materialist interpretation.

Pusey, for instance, was not alone in concluding that Darwin was unwittingly introducing the notion of "a power acting according to design." Samuel Wilberforce too had sensed that Darwin was implicitly ascribing to Nature the same ontological status as his fellow theists customarily ascribed to God. It should perhaps be pointed out straightaway that "Soapy Sam" Wilberforce was not the surpliced buffoon of later parody. He was an Oxford First in Mathematics who also had practical experience of livestock breeding and could rely for specialist information on his personal friendship with Britain's top paleobiologist, Richard Owen. His review of the *Origin*, which even Darwin regarded as "uncommonly clever" for its highlighting of weak points in his argumentation, ran to forty pages.[6]

Darwin's apparent raising of external nature to crypto-divine status was, concluded Wilberforce, just as much an article of faith as any of the more conventional forms of theistic belief. Wilberforce surely had a point. As Dov Ospovat ascertained in a foundational study some four decades ago, Darwin was often influenced by theological ideas without always recognizing them as theological ideas.[7] Ospovat noted significant similarities of outlook between Darwin and other naturalists of his era, including the legendary German explorer Alexander von Humboldt (1769–1859). Darwin took with him aboard the *Beagle* Humboldt's account of his own explorations, and from Humboldt he is thought to have derived ideas in the German tradition of *Naturphilosophie*, which includes the conception that all nature is governed by a "world spirit."[8] Ospovat also found that in his notebooks Darwin made reference to the fact that the creator creates by what Darwin elsewhere terms designed laws.[9] These, Ospovat says, represented "a far higher conception of the creator than to suppose he created by separate interpositions of divine power."[10]

The Humboldtian influence on Darwin's thinking may go some way to explaining why Darwin was so confident that the transmutation of species would reliably result in physiological improvement and development rather than accidental degradation. Yes, the idea of natural selection weeding out the defects had something to do with this, as did the geological column, suggesting as it did at least a very rough upward trajectory from simpler to more complex. But there seems to have been something else at work as well: a belief in something like crypto-providence, a faith not difficult to see as springing from the Christian cultural milieu of his upbringing. He does not seem to have questioned why a process he insisted was blind should somehow be automatically in favor of progress. Indeed, even an appeal to a progressively advancing fossil record can be overstated. After all, the same fossil record that begins with microbes and eventually reaches primates and eventually humans is also marked by what appears to be a series of cataclysmic global extinction events. The fact that natural selection was effectively a theory of progressive *development* as much as of random or merely opportunistic adaptation appears to have been a matter of unexamined faith for him.

What has felicitously been termed Darwin's epistemological double vision[11] may at first seem curious, and yet it was familiar enough to the mid-Victorian mindset, having been shared by a writer famed for her agnostic views, George Eliot, of whom it has been written that she "abandoned a deeply-felt form of Christianity, but retained in her agnostic years a deep sense of divine providence in the world."[12] The unresolved tension, which Darwin shared with many Victorian coevals, was something which began in his youth and endured to his older years. This can be witnessed if we make a short digression back to 1833 when a young Charles Darwin was making arrangements to board the *Beagle* for his global voyage of exploration.

Darwin's *Beagle* Library

In addition to Humboldt's account of his explorations, Darwin also brought the first volume of Charles Lyell's *Principles of Geology* trilogy (1830–1833) with him on the *Beagle* voyage. Lyell's ideas of geological uniformitarianism were to provide the intellectual foundation for Darwin's theory of biological gradualism-cum-natural selection.

Another book he chose to bring may strike some as out of left field: John Milton's seventeenth-century epic masterpiece, *Paradise Lost*. Lest one imagine the book was foisted on him by, say, his devout mother, note that he confessed to having favored that epic over all other reading matter he took aboard with him.[13] Darwin also informs us that his seafaring companions were apt to rib Darwin for coming across as a trifle pious in his frequent recourse to scripture to prove this or that point of principle being debated aboard ship. So did Darwin regard the twelve books of Milton's extensive work as a form of devotional reading? Seen in that light it might seem only natural for Darwin to carry with him his dog-eared Milton volume, but I doubt this was the whole story.

Paradise Lost has often been regarded in the English-speaking world as a grand poetic complement to the Christian master narrative contained in the Bible itself; yet critics down through the years have also noted that the epic is a mightily ambiguous exemplar of devotional literature. Though the work explicitly claims to "justify the ways of God to men," some have read it as doing precisely the opposite. In the last two centuries a few critics have read it as, if anything, a *theodicy manqué* rather than as an unexceptional demonstration-in-verse of God's beneficence to humankind. To be sure, Milton begins his work conventionally enough:

> What in me is dark
> Illumin, what is low, raise and support;
> That to the highth of this great Argument
> I may assert Eternal Providence,
> And justifie the wayes of God to men.[14]

Yet the heroic depictions of Satan as the great military commander of the fallen angels has struck many as being suspiciously reminiscent of that peerless hero of the classical tradition, Prometheus, in his opposition to the tyrannical and arbitrary gods of Mount Olympus. Milton's Satan pits himself against what he regards as divine despotism. God is, on Satan's view, "our grand Foe, / Who now triumphs, and in th' excess of joy / Sole reigning, holds the tyranny of Heav'n".[15] This conviction provides the motive for Satan's defection from heaven, illustrated in his address to the other fallen angels where he describes himself as the bringer of freedom: "I come no enemy, but to set free / From out this dark and dismal house of pain".[16]

I have cited one or two of Milton's purple patches of revolutionary rhetoric to help show why literary critic William Empson, in a famous study, described Satan as the de facto hero of an epic of cosmic revolt against divine repression.[17] The atheist Empson went on to write of the internal tension which must be experienced, he claimed, by all orthodox Christians on reading the epic, claiming that anybody not "imprisoned by their own [Christian] propaganda" would be impelled to see Milton's God as "monstrously wicked." This was a view which had in fact already been foreshadowed in the poet William Blake's famous contention that Milton had been of the devil's party without knowing it.[18]

There may well have been a biographical element in Milton's revolutionary eloquence. He himself had rebelled against the doctrine of the divine right of kings, had rejoiced in the execution of the absolutist Charles I, and had welcomed the subsequent advent of a parliamentary democracy. Later he deprecated the restoration to the throne of Charles II. The historical record shows that disputes over the royal prerogative (embodied in Charles I's defiant defense of the doctrine of the divine right of kings against the democratically mandated demands of parliament) led to the civil wars of the 1640s in England and eventually to the subsequent execution of the king. The later restoration of the Stuart monarchy with the enthronement of Charles II in 1660 symbolized for Milton the English nation's tragic rejection of its God-given right to republican freedom. *Paradise Lost* (1667) was written in the aftermath of this Stuart restoration to the throne and it is likely that republican sentiments lay behind Milton's strikingly idiosyncratic treatment of his epic subject matter.

Aspects of Milton's own political disappointment can be discerned behind some of the rebellious sentiments of his Satan figure. The fact that Satan gets the best lines (God's words come over as legalistic and uninspiring) may imply a barely concealed theological subtext, Milton charging God with dereliction given recent political developments—developments which, given Milton's antipathy for the anti-democratic "divine right of kings" doctrine, he can have viewed only as the very antithesis of providential.

A telling, parallel situation was recorded by the celebrated diarist Samuel Pepys, who lived through the same troubled times as Milton. Pepys and his employer and patron, Edward Montagu, the Earl of Sandwich, both harbored republican dreams. After the Restoration Pepys reported in his diary a perceptible diminuation of faith on Montagu's part.[19] As Pepys's biographer, Claire Tomalin, noted, "The Earl's disavowal of the faith of his youth was understandable, since for him and his former [anti-monarchist] party, religion had failed."[20] With these historical factors in mind, the intriguing question arises as to whether Darwin could in part have chosen *Paradise Lost* as reading material because he found his own ambiguous attitude to the divine reflected in the unmatched power and passion of Milton's verse.

This is not of course to deny that Milton can be enjoyed innocently as "pure" poetry of the art-for-art's sake kind. The late seventeenth-century writer Joseph Addison would later complain of Milton that "he writ no English," which is in a sense perfectly true because Milton created a higher, transfigured idiom all his own with the capacity to bewitch generations, the result being that Addison's complaint is now often glossed as a de facto compliment. Milton in fact wrote what present-day linguists would term a "superposed" variety of English bristling with Latinisms and other learned exoticisms which cumulatively reach such a pitch of stately majesty as to be reminiscent of church organ music at its most sonorous. It is more than possible to lose oneself in the purely aesthetic dimension of Milton's craft.

Nevertheless, despite the intrinsic beauties of what critics refer to as the Miltonic sublime, it is difficult to imagine Darwin reading *Paradise Lost* for its poetic cadences alone. For it seems more than merely coincidental that Darwin's own wrestlings with his faith mesh so closely with that theological subtext which critics from Blake to Empson have discerned in Milton's epic. That fraught and "conflicted" attitude to religion is reflected at many points of Darwin's correspondence with intimates, nowhere more illuminatingly than in a letter he wrote to supporter and theistic evolutionist Asa Gray May 22, 1860:

> With respect to the theological view of the question; this is always painful to me.— I am bewildered.— I had no intention to write atheistically. But I own that I cannot see, as plainly as others do, & as I sh^d wish to do, evidence of design & beneficence on all sides of us. There seems to me too much misery in the world. I cannot persuade myself that a beneficent & omnipotent God would have designedly created the Ichneumonidæ with the express intention of their feeding within the living bodies of caterpillars, or that a cat should play with mice. Not believing this, I see no necessity in the belief that the eye was expressly designed. On the other hand I cannot anyhow be contented to view this wonderful universe & especially the nature of man, & to conclude that everything is the result of brute force. I am inclined to look at everything as resulting from designed laws, with the details, whether good or bad, left to the working out of what we may call chance. Not that this notion at all satisfies me. I feel most deeply that the whole subject is too profound for the human intellect. A dog might as well speculate on the mind of Newton.— Let each man hope & believe what he can.—
>
> Certainly I agree with you that my views are not at all necessarily atheistical. The lightning kills a man, whether a good one or bad one, owing to the excessively complex action of natural laws,—a child (who may turn out an idiot) is born by action of even more complex laws,—and I can see no reason, why a man, or other animal, may not have been aboriginally produced by other laws; & that all these laws may have been expressly designed by an omniscient creator, who foresaw every future event & consequence.

But the more I think the more bewildered I become; as indeed I have probably shown by this letter.[21]

Darwin appears no less bewildered in the latter part of his life, as illustrated by two of his letters. In the famous letter to his botanist friend Joseph Hooker (February 1, 1871), he counters the view that with the recent demise of spontaneous generation as a viable scientific paradigm, the idea of a purely materialistic origin of the first life has also been put on its back heels. The thinking was that the conditions at the time of the origin of the first life are not so different from today's, and that it had recently become clear that life does not spring out of non-life in the present, so why should we expect it to have done so in the distant past for the origin of the first life? Darwin demurs:

> It is often said that all the conditions for the first production of a living organism are now present, which could ever have been present – But if (& oh what a big if) we could conceive in some warm little pond with all sorts and ammonia and phosphoric salts,—light, heat, electricity etc., present, that a protein compound was chemically formed, ready to undergo still more complex changes, at the present day such matter would be instantly devoured, or absorbed, which would not have been the case before living creatures were formed.[22]

So here again we have the Darwin bent on explaining the history of life without reference to any direct divine action. On the other end of this same decade, in 1879, he would write to another correspondent:

> It seems to me absurd to doubt that a man may be an ardent Theist & an evolutionist.…. What my own views may be is a question of no consequence to any one except myself.— But as you ask, I may state that my judgment often fluctuates. Moreover whether a man deserves to be called a theist depends on the definition of the term: which is much too large a subject for a note. In my most extreme fluctuations I have never been an atheist in the sense of denying the existence of a God.— I think that generally (& more and more so as I grow older) but not always, that an agnostic would be the most correct description of my state of mind.[23]

But then in his autobiography, penned primarily in 1876 but with additions and revisions over the ensuing five years, he noted that at the time of writing the *Origin*, he experienced "the extreme difficulty or rather impossibility of conceiving this immense and wonderful universe, including man with his capacity of looking far backwards and far into futurity, as the result of blind chance or necessity. When thus reflecting I feel compelled to look to a First Cause having an intelligent mind in some degree analogous to that of man; and I deserve to be called a Theist."[24] Notice the use of the present tense there. But note, too, that in this same autobiography he confessed that this sentiment "has very gradually with many fluctuations become weaker"[25] and that "disbelief crept over me at a very slow rate" and "was at last complete."[26] The last of these snippets, it should be emphasized, occurs five pages before he stated, present tense, "I deserve to be called a Theist"!

After all that wavering, he takes a stab at something approaching a definitive statement of his position by suggesting that any difficulty in throwing off belief could be due to childhood indoctrination, and finally he writes, "I for one must be content to remain an Agnostic."[27]

It is difficult (some would say, impossible) to postulate at one and the same time that sentient beings are the result of divine creation *and* of a random chemical reaction, and this tension underscores Darwin's abiding ambivalence on this subject. Darwin comes closest to managing the trick when he retreats to a tentative endorsement of a deistic stance in which all the divine creativity is somehow front-loaded into the laws of nature, at the earliest moment of creation, with the details of how such a front-loading trick could possibly be achieved left to the imagination. What is clear is that Darwin wants nothing to do with any outlook involving a creator actively creating in the history of life.

And that was but the half of his qualms with how the Bible depicts the Creator. In his autobiography he writes that the Old Testament, which presents God as a "revengeful tyrant," can be no more trusted than the "sacred books of the Hindoos, or the beliefs of any barbarian."[28] The doctrine of everlasting punishment in hell, Darwin continues, was a "damnable doctrine"[29] which would condemn his father, brother, and many of his friends. Because of this, he could "hardly see how anyone ought to wish Christianity to be true."[30]

His autobiography also addresses the theological problem of earthly suffering, elaborating on a theme he broached in his 1860 letter to Asa Gray:

> That there is much suffering in the world no one disputes. Some have attempted to explain this in reference to man by imagining that it serves for his moral improvement. But the number of men in the world is as nothing compared with all the other sentient beings, and these often suffer greatly without any moral improvement. A being so powerful and so full of knowledge as a God who could create the universe, is to our finite minds omnipotent and omniscient, and it revolts our understanding to suppose that his benevolence is not unbounded, for what advantage can there be in the sufferings of millions of the lower animals throughout almost all endless time? This very old argument from the existence of suffering against the existence of an intelligent first cause seems to me a strong one; whereas, as just remarked, the presence of much suffering agrees well with the view that all organic beings have been developed through variation and natural selection.[31]

Those words contain a whole tangle of theological assumptions and implications. On the one hand there is of course the tone of theomachy, that is, battling against the gods, familiar from the classical tradition of Lucretius and Epicurus.[32] Those two ancient authors felt that the morally compromised gods of the classical pantheon made them unworthy of veneration. Darwin expresses a similar sentiment here, at first taking what C. S. Lewis would later call the problem of pain as a sufficient reason to disqualify God from respectful consideration, and then even to register some doubt as to His bare existence. On the other hand, and in contrast to that tone but consistent with his divided state of mind regarding spiritual matters, the reproach against God (parsing the above words more minutely) gives way to an exoneration of the Divine. Darwin's words may be glossed as an implicit form of theodicy in that he seems to suggest that God, on Darwin's evolutionary take, cannot be held responsible for the imperfections of the human/animal estate. Instead, all the pain and evil are laid at the feet of evolution by natural selection. In other words, for Darwin, his theory provided an exculpation for God, letting Him off the hook for the evils of creation.

There emerges from all this the picture of a man of two minds, that kind of irresolution which modern psychologists, with their penchant for decanting old wine into new bottles, label cognitive dissonance. It was a state of mind to which Darwin was more than usually prone, his questioning and questing habit of mind standing out in sharp relief against the dogmatic certitude of some members of today's biological guild.

To be sure, Darwin's writings show him *fencing with* God, but this itself is testimony that he never lost the sense of God as a real presence. This divided aspect of Darwin's mind, although often a torment to him, lends the man an interest and complexity either over-looked or else passed over by modern, more doctrinaire exegetes of his intellectual legacy. It may go a considerable way to explaining Samuel Wilberforce's suspicion that Darwin's purportedly dispassionate science was a covert form of natural theology, an impetus not fully apprehended by Darwin himself.

Writing as Confessional: Metaphorical Disclosures

Wilberforce's assessment of Darwin is supported by close textual analyses of some of Darwin's words, where we are afforded glimpses of his state of mind from the metaphorical language he favored. Since it may strike some as a mere distraction to focus on his metaphorical language, as some may view such language not as the meat of Darwin's argument but instead as a mere crutch to non-scientists struggling with the technical material, it is worth pausing to consider more carefully the role of metaphor in scientific writing and how such metaphorical language has been regarded in the age of science.

In 1666, Royal Society member Samuel Parker dismissed metaphorical language as "wanton and luxuriant phantasies climbing up into the Bed of Reason," figures of speech that "do not only defile it by unchast and illegitimate Embraces, but instead of real conceptions and notices of things impregnate the mind with nothing but Ayerie and Subventaneous Phantasmes."[33] The irony, of course, is that this bracing dismissal of metaphor gains much of its force through the use of metaphor. The reality is that language, even reasoned discourse, is shot through with metaphor. Indeed, it's difficult to find a seemingly straightforward word that does not have a metaphor somewhere in the misty reaches of its past, lurking like some long-forgotten horse thief far back in one's family tree. Conceptual interferences arising from metaphors, distant analogies, and widely dispersed narrative patterns with deep roots in people's imaginations have long been discussed across the whole range of human cultures.[34]

Even in the context of supposedly objective scientific reporting, Misia Landau detected surprising interferences from folklore and myth,[35] warning that scientists should be aware of the power of pre-existent narrative structures to exert a subconscious influence on the way they present their would-be objective data.

In a similar vein, Andrew Reynolds recently drew attention to the sizable role played by analogical reasoning in Darwin's thinking. Such reasoning was "reliant on several key metaphors," he writes. "One was the Tree of Life to represent the thesis of the community of descent or shared ancestry of all species. The other concerned his hypothesized mechanism for species transmutation, which he called 'natural selection,' a choice of terminology based on an analogy with the process of artificial selection practiced by humans in the production of new varieties of domesticated plants and animals."[36]

As noted above, against the objections of numerous colleagues who underscored the problem with equating how mindless nature acted and what animal breeders did purposefully by the use of human ingenuity, Darwin initially persisted in claiming a close analogy between the artificial breeding methods of such persons as pigeon-fanciers and the claimed "selection" performed by nature. He was explicit about this claim, stating that he favored the term natural selection in order to mark its relation to man's power of selection.[37]

Perhaps drawing auxiliary strength from ancient ideas of an active and directive Nature (a thoughtway now lost to us but nevertheless an ancient conception which achieved a kind of swansong in the middle of the nineteenth century[38]), Darwin proposed that Nature, with limitless millennia at her disposal, could do a more comprehensive job of bringing about major physiological changes and new species than could human breeders. Yet as fond as he was of the analogy, he would eventually walk away from it, as becomes clear from close attention to his evolving choice of metaphors and other descriptors in discussing natural selection.

Darwin as Writer

On the face of it, Darwin's *Origin* might seem to be an unlikely quarry for the attentions of literary critics. With his habitual modesty, Darwin was commendably frank about his linguistic infelicities and the multiple difficulties he and wife Emma faced knocking the unwieldy volume into publishable shape. There must even have been some doubt whether the book would have appeared at all in 1859 had it not been for the help provided by his unstinting wife, his daughter Henrietta, and Emma's friend Georgina Tollet, who at different stages in the book's development helped with copying (Darwin's handwriting was notoriously difficult) and ensuring that spelling and grammar passed muster.[39]

Despite all the infelicities, however, there is no mistaking Darwin's enthusiasm for a subject which consumed his waking life. He was possessed by the mysteries of life and evolution, and something of the wonder Darwin felt when presenting his findings and expounding his ideas can be caught in a host of metaphors and rhetorical purple passages, many of which also suggest some rather arresting subtextual meanings and latent implications.

A particularly sustained effort in this vein is Gillian Beer's *Darwin's Plots: Evolutionary Narrative in Darwin, George Eliot and Nineteenth-Century Fiction*. There she makes the surely uncontroversial point that *any* important text is best submitted to close reading. "One's relationship to ideas depends significantly on whether one has read the works which formulate them," she comments. "Ideas pass more rapidly into the state of assumptions when they are *unread*. Reading is an essentially question-raising procedure."[40]

Beer's recommendation of best practice here was enthusiastically taken up in a more recent volume by critic George Levine in which, although he sometimes overdoes the ascription of literary virtuosity,[41] does have some arresting insights into Darwin's conceptions as they spring from the author's verbal and stylistic choices.

Levine, in line with Beer, insists that "coming to terms with what Darwin really said entails coming to terms with the *way* he developed his ideas and the language he found to express them."[42] Levine's discussion of Darwin's central metaphor, natural selection, is particularly fruitful. When questioned by colleagues, Darwin was sometimes inclined to dispute or pooh-pooh the implications of his own metaphors, but Levine's analysis of the verbal texture of Darwin's argument disallows such evasions. Metaphors are not, he contends, to be "skimmed off," for they play an important role in conveying the totality of the message any given author wishes to convey. As Stephen Jay Gould (no mean English stylist in his own right) once observed, "Our mind works largely by metaphor and comparison, not always (or often) by relentless logic."[43]

Nature Divinized

Darwin's metaphorical thinking is sometimes subject to a tension between nature as destroyer and nature as nurturer. Hard questions from one or more friends, and from reviewers of the book, led him to consider replacing talk of "natural selection" with the term "natural preservation," so as to avoid errant associations with choice and, by extension, teleology.[44] But Darwin seems to have sensed that such acquiescence would have been fatal to his idea of major morphological change. After all, his book was about the *origin* of new species, not their *preservation* having once arrived. Darwin needed the term "natural selection" to modify the reader's conception of nature's role— not an exterminator or even a mere embalmer of the status quo but a nurturing facilitator.

Hence we encounter Darwinian locutions such as natural selection "tending" to its innumerable charges, or "daily and hourly scrutinizing" in order to foster healthy developmental outcomes. Essentially, we witness Darwin's switching of metaphorical allegiance from one historically attested conceptual archetype to another, from the idea of nature as blind, uncaring, and chance-driven to that of Nature as a maternal figure.

For these reasons it is tempting to sympathize with Levine in his bracing contention that Darwin in the back of his mind conceived of natural selection not as an indifferent process but as "a woman, perhaps a goddess"[45] whose operations have every appearance of working teleologically. Her sway seems in fact to be so positive, even providential, that Robert J. Richards once termed Darwin's conception of natural selection a form of "divine surrogate."[46] Indeed, when parsed carefully, the lexical and metaphorical structure of Darwin's argumentation emerges as little less than a veiled description of the workings of something like the medieval goddess Natura (or Physis in her Greek form) newly rehabilitated for modernity.

Given such a subtext, it is little wonder that Darwin held out so long against Alfred Russel Wallace's argument that natural selection and human breeding methods were about as comparable as chalk and cheese. For the unacknowledged subtext of Darwin's presentation of natural selection is that it came close in his mind to becoming the de facto goddess of the natural world. This would explain his enormous faith in what he stated were the *directive* powers of a process which others could see only as unfathomable and wholly unpredictable (such having been the original connotation of the term "natural selection" coined by breeders).

Judging from a number of Darwin's evocations, the powers of natural selection transcend human intelligence to such a degree that he came perilously close to imputing to it the capacity for intelligent design, as observed above. The capstone to this crypto-theological way of thinking appears to have been laid when he contrasted the selfishness of mankind with the dispassionate care for *all* animal and human life shown by natural selection. Here natural selection seems

curiously reminiscent of the limitless goodness of the Christian God, while being contrasted with the sinfulness of mankind.

Levine, like many others before him, repeats the point that "it has long been understood that Darwin was much influenced by the [Paleyan] Natural Theology that much of the *Origin* is devoted to displacing" and that "his imagination of nature is of a designed place, and he adapts many natural theological terms, not least 'adaptation' and 'contrivance' in his description of the way nature works."[47] Levine's response to the text in his capacity as a literary critic leads him to corroborate Richards's view that Darwin did indeed have a divinized conception of nature. As Jerry Fodor and Massimo Piattelli-Palmarini have trenchantly observed, Darwin strove to exorcize "ghosts in the machine" such as God, selfish genes, or a World Spirit, yet "Mother Nature and other pseudo-agents got away scot-free."[48]

Darwin would defensively counter such accusations by insisting that in using the term natural selection, he had meant only the aggregate action and product of many laws.[49] However, this denial founders on his inability to account for the meanings disclosed by the metaphorical terms in which he clothed his ideas. Those more oblique modes of self-disclosure provide hints of the way his mind was working, partly at a sub-rational level, what in modern parlance might be called "tells." In sum, the linguistic analysis above corroborates the point first made by Edward Pusey and later endorsed by others, to the effect that the Darwinian conception of nature was, for all his protests to the contrary, that of a power acting according to design.

I have already more than hinted that Darwin was not romancing idiosyncratically. Instead, his imagination was nourished by conceptual and imaginative templates he had access to in early nineteenth-century culture. But to properly recover and explore these, some old-style scholarship is required. This will allow us to re-enter a thought-world which was Darwin's birthright and came to him as easily as breathing. An important clue on our journey there was recently provided by Ronald Hutton in his *Queens of the Wild: Pagan Goddesses in Christian Europe*,[50] to whose chapter on Mother Earth we now turn.

The Eternal Mother Figure

The tradition of a personified *Natura creatrix* (Nature the creator) has a long pedigree in Christian and pre-Christian European thought, stretching back at least to the ancient Greek Physis and her Roman counterpart, Natura.[51] To Aristotle, Physis was the force which generated and animated living things and embodied the elements and primary materials of the world. Aristotle saw her as a form of governess and steward to the sublunary world—as opposed to that of the heavens which were envisaged as the place of the most supreme divinity. She was sometimes assimilated to Plato's conception of the anima mundi (world soul or spirit), an entity endowed by the creator god with the role of linking his ideal realm with the realm of material and mortal beings. Some poets of the imperial period (such as Ovid) conceived of her as a cosmic power subordinate only to the Creator himself. The later writer Claudian referred to her as Mother Nature and credited her with producing an ordered world out of chaos.

Natura was inducted with some ease into the Christianized world in the fifth century, and in the twelfth century the philosopher and poet Bernard Silvestris portrayed her as having been created by the Christian God, tasked by Him to put the finishing touches to His universe. Similar conceptions were expressed in the work of Alan of Lille towards the end of the twelfth century when he too saluted Natura as the ruler of the world on behalf of the Christian God. Jean de Meun, in the *Romance of the Rose*, referred to her as God's chamberlain, and Chaucer depicts her essentially as God's deputy.

In sum, concludes Hutton, "the concept of a mighty female figure embodying and ruling over the terrestrial world was embedded in Christian intellectual and literary culture all through the periods in which Christianity most completely dominated Europe, the medieval and early modern,"[52] that is, up until about 1650. One may debate to what degree such language was regarded as literal versus metaphorical in European society and for any given individual, but what is undeniable is that she continued her reign up to the nineteenth century, maintaining a secure place in the Christian cosmological imagination.

The Mythic Universal

Turning towards Darwin's own day, in the later Romantic period of the nineteenth century, when poets like William Wordsworth were exhorting readers to let nature be their teacher, Natura was eulogized as a fount of wisdom. The poet Percy Bysshe Shelley even apostrophized her as "Sacred Goddess, Mother Earth/Thou from whom whose immortal bosom/Gods and men and beasts have birth." The poet Algernon Charles Swinburne too conceived of Nature as a mighty female deity, embodying and creating the universe itself. Hutton also points to the remarkable example of Charlotte Brontë's *Jane Eyre*, whose eponymous heroine, finding herself one night alone and sleeping rough on a moor, is comforted by the thought of Nature, conceived of as a maternal figure, and by that of a loving God, as Nature's creator.

The idea will seem alien to many of us, but it was in such filiations that Natura remained, as Hutton puts it, "resiliently adaptable" up to Darwin's day and slightly beyond. Over countless centuries there had established itself in Europe a "Christian pattern of a cosmic feminine force subject to a patriarchal sky deity."[53] Some modern archaeologists have even gone so far as to claim that the archetype of the Great Mother has been a mythic universal—an image inherent in or somehow "hard-wired" into the human psyche since prehistoric times.

Such historical representations of Natura and the fact that she continued to be so deeply embedded in Victorian cultural understanding make it likely that the idea was lodged in Darwin's psyche too at some level and that his conception of natural selection may not have been anywhere near as purely materialist as the strictly Cartesian part of his mind might have wished. Despite his rational(ist) efforts, it seems he was never able to free himself completely from that form of Christian cosmology which enjoyed such wide currency in the ambient culture of his day.

Prosaically summarized at the purely rational level, the *Origin* aspires to supply a fresh, materialist myth to account for the development of the earth's numerous species. In reality, however, that single-minded project is compromised or even subverted by an insistent metaphorical subtext that introduces a covertly metaphysical dimension into Darwin's conceptions. That submerged stratum of meaning, with its unmistakable echoes of the figure of Natura in Christian cosmological thought, makes its influence felt at many metaphorical and lexical levels.

It was not long before Darwin's peers called him out on this. In the end Darwin was persuaded to capitulate on multiple fronts, as for instance when he conceded, in a September 1860 letter to Charles Lyell, "Talking of 'Natural Selection,' if I had to commence de novo, I would have used natural preservation."[54]

This emendation amounted to much more than mere verbal finessing: Darwin's self-correction brought with it significant conceptual consequences. The letter to Lyell involved a fatal concession which, had it been made public and analyzed dispassionately at the time, could have halted the unstoppable march of Darwinism there and then. For as noted, the term that in September 1860 Darwin conceded was preferable, "natural preservation," by definition refers to a process that only preserves rather than creates new body parts, let alone whole new body plans. The whole Darwinian theory of an advance from organic simplicity to complexity—from microbes to man—must at best fall moot after such a major semantic retreat.

Even Darwin himself seemed to harbor doubts. Could natural selection *really* have exerted the vast transformative powers he had claimed for it? Such doubts could account for his flirtation with what we might call supplementary Lamarckism—present in the first edition of the *Origin* and to an increasing degree in subsequent editions. His doubts also might go some way to explain the otherwise anomalous peroration in the *Origin* to the effect that evolution had come about by dint of "laws impressed upon matter by the Creator."[55]

One might even speculate that Darwin's two-decade procrastination over publication of the *Origin* might have owed something to his difficulties in convincing *himself* of many ideas which, on the advice of colleagues and critics, he was driven to modify quite considerably over his five later revisions of the book. Most strikingly, the selection/preservation problem would appear to have anticipated some significant modern developments in biological research, a brief account of which will now be intercalated. As will be seen, modern developments have provided a retrospective confirmation of some of Darwin's worst fears.

A Modern Postscript: The Evolutionary Black Hole

As G. B. Müller and S. A. Newman point out, population genetics, and thus evolutionary biology, has not identified a specifically causal explanation for the origin of true morphological novelty during the history of life.[56] In other words, neo-Darwinism simply cannot account for *nontrivial* innovations. Darwin's rowing back on that point to Lyell was, whether he fully acknowledged it or not, fatal to any macromutational claims. As leading biochemist Nick Lane explains, "It is generally assumed that once simple life has emerged, it gradually evolves into more complex forms, given the right conditions. But that's not what happens on Earth.... If simple cells had slowly evolved into more complex ones over billions of years, all kinds of intermediate forms would have existed and some still should. But there are none. Instead, there is a great gulf."[57] Or as he says elsewhere, "There are no surviving evolutionary intermediates, no 'missing links' to give any indication of how or why these [cellular] complex traits evolved, just an unexplained void between the morphological simplicity of bacteria and the awesome complexity of everything else. An evolutionary black hole."[58]

What is the upshot of all this? "This means there is no inevitable evolutionary trajectory from simple to complex life," Lane insists. "Never-ending natural selection, operating on infinite populations of bacteria over millions of years, may never give rise to complexity. Bacteria simply do not have the right architecture."[59] A striking admission from a mainstream evolutionist.

But, of course, speciation did occur, and there was a general trend from less complex billions of years ago to more complex, even if the fossil record suggests that the increases in complexity arrived in sudden and tremendous leaps rather than in the agonizingly gradual fashion

called for by Darwinian theory. So how were all these novel biological forms and increased complexity managed? To save appearances,
competent scientists are thrown back on the placeholder terms fate/
chance, such being all too plainly a cover for ignorance.[60] Fodor and
Piattelli-Palmarini are more refreshingly candid when they pose the
rhetorical question: "So if Darwin got it wrong, what do you guys think
is the mechanism of evolution?" Their answer: "We don't know what
the mechanism of evolution is." Elsewhere in the same work they add,
"Nor did Darwin, and nor (as far as we can tell) does anybody else."[61]

Confusion Twice Confounded

The tenability of natural selection as a coherent postulate encountered
another bump in the road even within Darwin's lifetime, and once
again the problem was rooted in language. This additional difficulty
revealed itself when it was decided to have the *Origin* translated into
German.

If an author is not completely clear about what he meant in pen-
ning a work, or unwittingly lends it hidden dimensions of meaning
he is not fully conscious of, then the task of translating it into another language can present acute difficulties. Darwin never made any
bones about his rather middling academic status at Shrewsbury School
(which he attended until the age of 16), and cheerfully admitted to
not being as bright as his own sister.[62] However, the difficulty with
Darwin's lapses in intellectual clarity and perhaps self-knowledge,
together with his having rather perversely turned the original, familiar
meaning of natural selection on its head, was to make itself felt acutely
in continental Europe when Heinrich Bronn was commissioned to
translate the *Origin* into German.[63]

Bronn, in company with almost all early reviewers, thought
Darwin's theory wrong[64] and added an addendum to his German
translation pointing out the weaknesses of Darwin's arguments, foremost among which was: Where are all the fossils claimed to document
Darwin's evolutionary links? Darwin was eventually to be so offended
by Bronn's objections[65] that he later moved to somebody he hoped
would be a "tamer" translator for a second German version, Victor

Carus; but both Carus and Bronn had the same trouble translating the key term "natural selection."

The trouble stemmed from the latter half of the term. Even English-speaking readers had been confused by the term: some had glossed it (not unreasonably, it may be thought) in the teleological sense of Nature making a conscious choice. Darwin himself had given some succor to that interpretation in numerous passages, one cited above, where he appears to describe Nature as actively choosing and winnowing. Bronn's initial choice of the term *Wahl der Lebensweise* (literally, "choice of life mode") was objected to by Darwin for being too purpose-driven and "Lamarckian."

Recall that the French naturalist Jean-Baptiste Lamarck had postulated a kind of self-help mode of evolution whereby a creature would positively adapt to its surroundings by dint of its own repeated choices. The giraffe, for instance, might over multiple generations have been able to elongate its neck by repeated exertions to browse for food at the tops of trees, the elongation becoming heritable over time. It did not have to wait passively for a rare genetic mutation that just *might* benefit it (but more than likely would harm it), as Darwin proposed. Unfortunately, as Darwin ally Thomas Huxley was wont to point out, beautiful theories all too often are crippled by brute facts. Lamarck's attractive speculation was one such theory. Advances in genetics in the early twentieth century showed that, to choose a modern example, a weightlifter's overdeveloped musculature is not a biologically heritable feature from which his male offspring could benefit far into the future. Although more recent findings have partially muddied the waters, the dominant view in biology has for many decades been that the Lamarckian theory does not present a viable biological pathway for macroevolution, certainly not in its unsophisticated original form and very probably not in any form.

In deference to Darwin's wishes, Bronn consented to delete Darwin's proscribed "W" word (*Wahl*, meaning choice or selection) and settled for the more noncommittal and rather vacuous term *natürliche Zuchtung* (natural breeding). That translation gives no inkling of what the mechanism behind the putative process could be, and so it is hardly surprising that Victor Carus later restored the offending *Wahl* word by translating natural selection as *natürliche Zuchtwahl* (natural *choice* in propagation). This, however, as was swiftly pointed out by a German contemporary, Ludwig Büchner, simply reinstated Bronn's original "error" by re-interpolating what Darwin saw (rightly or wrongly) as the translation's teleological error.

Büchner pointed out that nature does not breed in the way of livestock farmers, explaining that in Darwin's way of thinking any selection must occur without purpose or intent—apparently passing over in tactful silence the blinding logical contradiction of "selection without intent" inherent in Darwin's thinking. Through no fault of their own, the two German translators found themselves caught up in an unholy semantic muddle when confronted by the "floating signi- fier" of natural selection—a term even Darwin himself was incapable of defining in satisfactorily precise and logical terms.

It should perhaps be noted at this point that the translators' dif- ficulties did not spring from any inadequacy of the German language to convey subtle English terms. No language is more adaptable and pliant than German; the claim that you can make up your own Ger- man words at will as you go along by improvising ad hoc compound nouns is not entirely a joke, underscored by the German word for compound noun, *Zusammensetzung*, itself a compound noun from two German words that mean "placing" and "together." The difficulty that the German translators had with the term was not merely *linguistic*; it was *conceptual*, and points to a far more fundamental incoherency in the term itself, and in Darwin's thinking.

German biologists had noted that among Darwin's writings he sometimes implied that there was a general plan or design for nature, which was not unreasonably glossed by some of his German colleagues as letting God in through the back door. That point was taken up by Eberhard Dennert in his polemically titled 1902 work *Am Sterbelager des Darwinismus* (*On the Deathbed of Darwinism*).[66] Dennert argued that evolution could in good logic only be a supranaturally directed process, since some originating force must have been instrumental in its unfolding. This point had in fact already been made in a particularly forceful way by Princeton theologian Charles Hodge in 1874 when he professed himself unable to understand how selection without any selective agency had occurred, prompting him to ask the devastating rhetorical question of whether natural selection was seriously to be understood as "a blind process of unintelligible, unconscious force, which knows no end and adopts no means."[67]

Hodge alleged that the term is so nebulous as to be meaningless— an "empty referent" in linguistic parlance. Some two decades after Hodge the eminent botanist William Bateson—shortly thereafter to be instrumental in pioneering the new science of Mendelian genetics in Cambridge at the beginning of the twentieth century—also queried the lack of substance in the term natural selection. Bateson pointed out that the vagueness of Darwin's description of natural selection as occurring by insensible and imperceptible stages gave us no clue as to what the precise operative mechanism might consist in.

More than a century after Bateson, we have learned much about the mechanisms of random variation (the other half of Darwin's formula), but as we saw from the sampling of comments from recent mainstream biologists above, the term natural selection continues to baffle.[68] The term, in short, remains something of a will-o'-the-wisp.

At the End of the Rainbow

The sciences, as it turns out, have a rich history of employing such signifiers, ones that are, if not wholly insubstantial, then at least every bit as slippery as an apparition.

There are terms we use in everyday life which are purely notional rather than factual. The pot of gold at the end of the rainbow and the man in the moon are two examples of what are sometimes jocularly referred to as "airy nothings." Such terms exist on paper but lack any substantial referent in the real world. Of course, nobody aside from children and the pathologically gullible are misled by such light-hearted *façons de parler*; the terms are widely understood to be only notional. However, there are cases, including in the sciences, where a term is taken for reality but turns out to have no more substance than the man in the moon or that ever-elusive pot of gold.

A famous example is the substance dubbed phlogiston, postulated in the 1660s as a fire-like substance inhering in all combustible bodies but proved not to exist just over a century later by the French scientist Antoine Lavoisier. In more recent times there is that apocryphal substance dubbed ectoplasm which was claimed by some Victorian spiritualists to denote a substance exuded from a "medium" thought to represent the materialization of a spiritual force once existing in a now-deceased human body.

Another example can be found in an imaginary dialogue penned by the seventeenth-century French writer Bernard de Fontenelle, in which a philosopher character expresses the belief that there very well could be a man on the moon—and a whole civilization to boot, if you please. In this 1686 work, *Entretiens sur la Pluralité des Mondes* (*Conversations Concerning the Plurality of Worlds*), Fontenelle presented an imaginary conversation in a country garden between a philosopher versed in astronomy and an enquiring hostess. The philosopher expresses confidence that the telescopes of the time, though primitive, appear to reveal a moon covered with seas, lakes, mountains, and valleys, even though, he concedes, it is not yet possible to observe lunar inhabitants. The similarity of the terrestrial and lunar landscapes, however, provides good reason for optimism that sentient life must exist on the Moon, the philosopher concludes.

The philosopher's wish has clearly become father to his thought. The imaginary notion has assumed concrete shape in his mind, a contention without empirical foundation or any locatable referent in the tangible world, but one nevertheless held to by the person who proposes it: in common parlance, a delusion.

Some modern scientists give the impression of being the intellectual descendants of Fontenelles's imaginative philosopher character in their endless quests to find evidence of life on irredeemably hostile planets. As late as 1905 the American astronomer Percival Lowell was still capable of succumbing to a subvariant of the man-in-the-moon myth by publishing a perfectly serious book on the subject of Mars and its supposed canals. Lowell followed in Fontenelle's footsteps by wrongly identifying planetary indentations as artificial constructions fashioned by alien denizens of that planet. There was however precious little to justify this belief, as Alfred Russel Wallace, by that time an octogenarian, pointed out when he came to weigh in on the controversy.

Wallace had already given powerful, scientifically grounded arguments against such a notion in his *Man's Place in the Universe* (1903) and later in *The World of Life* (1910), where he deployed his technical knowledge about the thinness of the Martian atmosphere and planet cooling ratios relative to distances from the sun, which, Wallace concluded, made Mars too cold to allow water to flow. Why build canals if there is no liquid water? (The lack of liquid water also makes it devilishly hard for nature to build Martians.[69])

This did not stop Lowell from advancing quite seriously his ideas on the artificial, presumably alien-built, canals. A key reason why he persisted in adhering to a faulty theory yet again boils down to the use of language, in particular the misunderstanding of the term *canali* first used by the Italian scientist Giovanni Schiaparelli to describe his observations of the Martian surface. *Canali*, in Italian, means conduits of some sort, whether natural or artificial. In English, however, "canals" have connotations of man-made excavations.

The above examples—phlogiston, Fontenelle's lunar civilization, and Lowell's Martian canals—are examples of the error well-known to linguists as a false friend (*faux ami*), cases where one is taken in by a concept present only in language, over against the testimony of the real world. In the most fortunate cases, the empirical testimony becomes so insistent that, as noted above, all but small children and the insane are awakened from the illusion. In the latter example, it was the images of the Mariner 4 probe in 1965 which put an end to the belief in intelligent Martian life. All previous postulations were now unmasked as being "just words" without real-world referents.

Non-Signifiers

As regards Darwinism, Fontenelle's lunar lakes and Lowell's Martian canals are not mere historical curiosities. Darwin's long shadow has, it seems, had the effect of licensing all manner of sub-scientific credulities. For instance, in the aftermath of the publication in 1976 of Richard Dawkins's *The Selfish Gene,* philosopher Mary Midgely pointed out, in painfully direct terms, the fatuousness of the Dawkins "meme" hypothesis and the whole notion of "cultural replicators."[70]

Her critique has since been reinforced by others. Philosopher David Stove[71] subjected the Dawkinsian conjecture to a particularly comprehensive critique and latterly John Gray, with his habitual trenchancy, has written of "the cod-science of memes"—those supposed replicating forces postulated by Dawkins which have no more referent in the real world, Gray fulminates, than the non-existent substance of phlogiston.[72]

This thrashing of Dawkins brings us back to evolutionary theory and the idea of evolution by natural selection. Philosopher Antony Flew, a longtime atheist and eleventh-hour defector from the ranks of the Darwinian faithful, came to doubt that the term natural selection had any substantial meaning, and he questioned the selective power Darwin claimed for it.[73] His British philosopher colleague Richard Spilsbury concurred, writing with devastating simplicity about the same term as well as that other question-begging mouthful "complexification." Concludes Spilsbury, "To say that these developments *might* have come about through the selection of chance variation is not evidence that they *did*."[74] To establish a conceptual possibility was far from advancing a concrete proof, he objected. To assent to that proposition was as futile as it would be to support Charles Darwin's long-since disproved theory of pangenesis and the associated idea of "gemmules" in heredity.[75]

From this chapter's brief linguistic survey, it appears to be all too easy to fashion words to elevate chimerical notions, the more convincingly if one chooses to append the honorific title of "science" to one's thought experiments. One can, for instance, muse with the ancient philosophers Epicurus, Lucretius, and the eighteenth-century Scot David Hume that the world evolved by chance collocations of atoms, but rather than stop there, take the additional rhetorical step of dignifying the notion by dubbing it "the theory of atomism." Similarly, one can submit for consideration that the universe and all within it arose spontaneously from entirely autonomous "natural law" by dint of a scientific-sounding process of "abiogenesis." The packaging has improved, but in all these cases we have to be willing to ignore that such theories involve what is known as "the suppression of the agent." This means the failure to specify who the agent/legislator might be. This is the sort of vagueness we were taught to avoid in school English lessons. One mundane example of this suppression is the criminal's perennial excuse, "The gun just went off in my hand, officer, honest."

It is then both grammatical solecism and logical impossibility to contend that the universe arose as an "agentless act" (Peter Atkins's term) since this would imply some form of pure automatism or magical instrumentality unknown to the material world. In an analogous vein one might, with Darwin, theorize that the development of the biosphere was simply down to that empirically unattested sub-variant of chance he chose to term natural selection. But given the rising tide of expert protest summarized above—first among biologists but also among philosophers—we are forced to consider the possibility that this term, like the others above, is one without referent, an empty signifier—or, at best, a highly exaggerated and misleading one.

Yet the urge to find proof for the godlike powers of natural selection endures, and in some quarters, not only to search for proof, but to insist, against a rising tide of evidence, that those powers are already proven beyond a reasonable doubt. We may take such behavior as a reminder that words, even airy nothings, have the power to create their own virtual realities in our minds, ones with little if any relation to a definable referent in the world of flesh, bone, and stone.

The bottom line for many present-day scientists appears to be, as mainstream evolutionary biologist Bob Holmes puts it, that "speciation still remains one of the biggest mysteries in evolutionary biology and that the unexamined view of natural selection leading to large-scale innovations is not true."[76] Even unreformed Darwin apologist Michael Ruse saw fit to note that natural selection cannot actually select anything and is better understood as a score-recording statistic than as the true cause of any particular phenomenon. "Natural selection is simply keeping score, as does the Dow Jones Average," he wrote. "The Dow Jones does not make things (cause things to) happen. It is just statistics about what did happen."[77]

Such concessions from mainstream evolutionists mean that important aspects of the Darwinian narrative once accepted as beyond question are no longer the solid pillars of scientific consensus they were once assumed to be. This is no small development. Darwin advanced natural selection precisely to rescue the idea of purely natural evolution of all life, that branch of speculation widely pooh-poohed before 1859 as a hobby-horse favored by armchair philosophers such as Charles's grandfather, Erasmus Darwin, and his French *confrères*. Charles hoped that the natural selection idea would revive the faltering program by supplying it with a semblance of quantifiable support. But now, with so many mainstream scientists publicly casting doubt on the power of natural selection, we are forced to ask ourselves: How could evolution in that more ambitious sense have taken place if the "true cause" thought to be driving it is now widely doubted by competent experts?

3. Piecing Together
a Theory

*The formation and transformation of Darwin's theory represented
not so much the results of an interaction between the creative
scientist and nature as between the scientist and socially constructed
conceptions of nature.... his interaction with nature was mediated by
assumptions and ways of perceiving nature that he derived from other
naturalists, both his predecessors and his contemporaries, and from
the culture in which he was educated and carried out his work.*
—Dov Ospovat[1]

I
N MANY RESPECTS DARWIN'S IDEAS RESEMBLED MORE A CUMULATIVE,
essentially social construction of reality than the result of straightforward, empirically demonstrable findings. He was, without fully
knowing or else acknowledging it, following in the footsteps of ancient
Greek and Roman natural philosophers who, innocent of the concepts
of empirical confirmation and experimentation, speculated freely on
the mainsprings of life and the universe with precious few checks
or controls on their imaginative forays beyond that of dialogue with
sympathetic interlocutors in the ancient worlds of Greece and Rome.

During recollections in his older age, Darwin looked back wist-
fully on his voyage aboard the *Beagle* in the 1830s as having been the
defining experience of his career, and one can easily see why. Those
five years had been the making of the younger Darwin, since they
rescued him from his previously ill-focused life. No longer a trial to
his long-suffering father, the son was able to devote himself to his

one true avocation in life as a gentleman naturalist. It is nevertheless necessary to sound a note of caution if one is tempted to regard that formative rite of passage as providing the foundation for his distinctive views on evolution.

To be sure, for understandable reasons to do with the aesthetics of constructing a compelling narrative, the Darwin legend has it that his South American experiences were responsible for the formulation of a host of evolutionary discoveries. According to the conventional, partly fictionalized story, the intrepid explorer returned from having garnered the secrets of nature in the exotic realms of the South Seas to share his secrets with his fellow men and women. Such a reading nicely fits the archetypal pattern of the heroic Prometheus who in Greek mythology brought down fire to earth from the abode of the gods in order to share it with mortal man. Resonating with people at a subconscious level, it is the kind of stirring story audiences like to hear, and reporters and other storytellers often have eager recourse to such archetypal narrative patterns because such patterns seem to be all but hard-wired into audience expectations of what a "proper" hero tale should consist in.[2]

As noted above, Misia Landau made an important contribution to this subject through her careful delineation of some surprising interferences in scientific papers from folklore and myth.[3] She recommended that scientists should be aware of age-old narrative structures since they could exert a subconscious influence on the presentation of modern data. Rather like the way we are tempted to embellish stories in everyday life to amuse interlocutors, she argued, the choice of narrative mode used to explain evidence can predispose the reporter towards traditional and readily intelligible patterns of understanding to the detriment of the true and unique particularities of the evidence being presented.

The Darwin legend certainly makes for a good story. But alas, once denuded of fictional and mythic accretions, one finds that the true story of the formation of Darwin's ideas of biological origins had little to do with romantic discoveries in exotic locales. Real life, as is all too often the case in human affairs, was more prosaic, with little resemblance to the finely honed romance narrative that was developed *around* Darwin.

In their study and edition of Darwin's account of his sea journeying and researches, Janet Browne and Michael Neve are firm on the point that Darwin's ideas did *not* come to him from his experiences in the field and that "the received image of Darwin voyaging alone through vast turbulent seas of thought as he paced the deck of the *Beagle* is a fantasy."[4] Darwin's evolutionary ideas did not derive from his empirical observations on islands of the equatorial Pacific or from any other region of the world. Rather, they emerged in a series of ad hoc, serendipitous installments, the result of his ability to weave together ideas culled from others, not all of them naturalists. To be sure, one might arrive at scientific truth by such a method, and there are notable scientists who have done so, or at least have begun in this way. But it's equally sure that many a phantasm has been chased by such a method, by scientists and layman alike, and we would do well to acknowledge that it was largely in this empirically far less rigorous mode that Darwin pursued his evolutionary theory.

The Family Tradition

In formulating his ideas Darwin was to derive inspiration from the most diverse sources. He later recalled that in the course of exploring the Galápagos Islands of the equatorial Pacific in the 1830s, he was much indebted to a purely chance remark of a colonial vice-governor, Nicholas Lawson, who met the *Beagle* when it arrived on his island and later, over dinner, claimed he could tell by merely looking at a tortoise from which island of the archipelago it had come.[5] In his autobiography, Darwin confessed that he, dedicated naturalist though he was, had not noticed these variations and their implications for evolutionary theory. Nor did he immediately follow up on the observation, not until a considerable time later when he was back in England.[6] But it was that drawing-room conversation with a colonial official on the island of Floreana that would eventually turn Darwin's thoughts in the direction of adaptive physiological change to differing environments and the potential for modifications and, eventually, transmutation of species. The official's small talk provided the bridge linking Erasmus Darwin's airy speculations about transmutation to his grandson's later ideas of speciation via random variation and natural selection.

The kinship of ideas between grandfather and grandson was something of an open secret in the nineteenth century. We might even imagine a partial recapitulation of the grandfather's intellectual life in the grandson's. Although Erasmus did not live long enough to become acquainted with those geological writings of Charles Lyell which so deeply influenced his grandson, he did read the similar views of James Hutton when they appeared in Hutton's earlier *Theory of the Earth* in 1788. The discipline of geology was to play an equally formative role in the thinking of both grandfather and grandson. Samuel Wilberforce showed full awareness of continuities in their thinking when, in the course of his lengthy review of the *Origin* he made the just observation, "If we go back two generations we find the ingenious grandsire of the author of *The Origin of Species* speculating on the same subject, and almost in the same manner with his more daring grandson."[7]

The family tradition of evolutionary speculation in fact had its origin even earlier than Erasmus's day. In 1719 Erasmus's father, Robert, brought to light the partial fossil of an ancient marine reptile described in the Philosophical Transactions of the Royal Society of that same year. It was later designated a plesiosaur fossil from the Jurassic and is today on display in the Natural History Museum in London. This find inspired Erasmus with the thought of common descent for the world's animal life, all life being thought to derive from a common ancestor in the form of a microscopic filament or "ens." The theory indicated that species changed and eventually died out over long ages, thus explaining the unearthed fossils of animals previously unknown to the eighteenth century.[8]

Intellectual Forebears

If one were to single out the most formative influence on Darwin the evolutionist, a plausible candidate for the part would be his older friend and unofficial mentor Sir Charles Lyell, author of the *Principles of Geology* (three volumes, 1830–33), the first volume of which was published in time for Darwin to take it with him aboard the *Beagle*. Many important aspects of the *Origin* have their intellectual roots in Lyell's *Principles of Geology*, which is not surprising since Lyell, building on the earlier *Theory of the Earth* by James Hutton, had laid out what was in essence an account of evolution as it applied to the non-organic world. Darwin, in essence, appropriated Lyell's principle of uniformitarian geological evolution and applied it to the world of life.

Lyell, of course, is but the tip of the iceberg. Darwin was a man of his age, as was his theory of evolution. The early Victorian age, the very crucible of the Industrial Revolution and of the modern scientific outlook, may seem familiar to us in many respects; but in terms of the metaphysical foundations of some of its thoughtways, it was truly a world apart from us. As theories of evolutionary gradualism were taking shape, so too was catastrophist theory. According to this framework, a series of catastrophes, including gigantic floods and seismic events, have been visited upon the earth, in each case bringing a life-extinguishing end to intervening periods of calm and stasis. All life was supposed to have ended at these catastrophic points, only to be restored. At each successive stage of renewal, life took a more advanced form in a supposedly vertical movement culminating at last in the creation of mankind. The French catastrophist Georges Cuvier (1769–1832) minimized speculation about the relation between these proposed catastrophes/renewals on the one hand and, on the other, the Genesis account and divine action. But in the British context, these catastrophes and renewals were readily attributed to divine action.

Catastrophism posited an extended geo-theological drama of death and rebirth spanning the eons. The framework's cosmic discontinuities, its phases of destruction and renewal, although grounded in the discontinuities of the fossil record, do have about them a mythological flavor. Think, for instance, of the Norse/Wagnerian myth of Ragnarök and the death and rebirth notions of Hindu philosophy. Such conceptions, arising independently in a number of cultures, supposed that life proceeded in a sequence of leaps forward following on from terminal reverses. That being the case, there could be no question of any phylogenetic continuity between the animals of one era and those of the succeeding era—which was of course the essential bedrock of Darwinian conceptions.

For those catastrophists not shy about invoking God in the process, all was said to have occurred in a series of divinely appointed stages dealt out with a savagely hands-on peremptoriness. This must surely have been a distinctly Old Testament God, and many found catastrophism difficult to square with ideas of Christian providence, thinking it more akin to the gods of the classical pantheon as reimagined by Shakespeare in the line "As flies to wanton boys are we to the gods / They kill us for their sport." Nevertheless, many people before the later 1830s were content to view the process in progressionist terms, successive catastrophes justified on the grounds that an ascending scale paved the way for the appearance of the glory of Creation, human life.

Against that schema for the history of life, Hutton and Lyell, focused primarily on inorganic geology, adduced purely natural causes such as might be still observable at the present day as the causes for all observable geological formations. This opposing theory of uniformitarianism—meaning the purely natural and non-divinely directed development of geological features over large tracts of deep time—won wide and swift acceptance, although not entirely without demur. The discontinuities in the fossil record persisted as a contrarian voice, and while a strategy of attributing all and sundry to a single great global flood had grown unfashionable in academic circles by the end of the nineteenth century, there has been a renewed openness to catastrophism in purely secular and geological terms. Present-day geologists and paleontologists reckon with several major mass extinctions over the ages, attributed to such natural phenomena as vulcanism and outsized meteor strikes.

Such events were allowed back in the door because they did not violate the most fundamental prejudice of the ascendant uniformitarianism, namely the commitment to banishing metaphysical speculations and supernatural causation from geology. Lyell, however, saw no need to extend this prejudice to the history of life. This became abundantly clear by the mid-1860s, when Lyell made common cause[9] with Alfred Russel Wallace, who had come to doubt the claimed far-reaching powers of chance variation and natural selection, concluding that human life at any rate must have arisen not by purely algorithmic processes of serendipity but through the ministrations of an intelligent initiator capable of forethought.[10]

In this regard, Darwin could be said to have been more Lyellian than Lyell himself. Darwin, whose ideas so often proceeded from analogical thinking, set his face against both natural and supernatural saltations (jumps, from Latin *saltus*) in the biological world—the notion that God had created various animal groups each in a single leap. After reading Lyell, Darwin began to see in biology and geology directly comparable patterns of gradualistic evolution. For him speciation—the development of different animal-types—had occurred over an extensive period of geological time in a process where originally simple organisms had modulated into increasingly complex forms.

At the opposite end of the spectrum from Darwin were those catastrophists who were essentially practicing science from a platform of biblical conviction, foregrounding what was known as the diluvial theory, centered on the biblical flood and the related story of Noah's ark recounted in the Old Testament. Lyell and others would refer to this form of thinking dismissively as Mosaic geology. This brand of catastrophism viewed the planet as having been molded by forces far more powerful than any observable at the present day; hence, the thinking went, these forces must have had a supernatural cause, set in train by God himself. In advancing the new uniformitarian and non-supernatural theory, Hutton and later Lyell signaled their dissatisfaction with such biblically-inspired geology—although Hutton had been prepared to entertain the ultimate possibility of a (divine) hidden hand behind all geological phenomena.

The upshot of the new thinking was to bring geology and theology into a potential collision course. Lyell, Anglican churchman though he was, wanted nothing to do with religious attempts to explain geological realities. As for the ultimate beginning of the physical universe, he adjudged that issue a metaphysical question and thus outside the proper business of empirical science. As Charles Coulston Gillispie noted in his classic treatise, Lyell built his work on the strict method-ological limitation that the past could be studied only by reference to what natural forces could accomplish at the present; anything beyond that would be unprovable and quasi-mystical.[11]

The upshot of Lyell's researches was to make it clear to all but biblical literalists that the world was not made in six days but rather evolved over countless millions of years, and that it was still evolving. Mountains, for instance, gradually subside over large tracts of time due to the slow, attritional effects of water erosion. This, modern geologists inform us, will be the long-term fate even of the mighty Alps. Our physical and topographical surroundings are dynamic, not static. Nothing is divinely fixed for all eternity.[12]

One can see why Lyell should have become popularly known as Darwin's John the Baptist. For Darwin's thinking appears to have been as follows: Since Lyell had removed the hand of God from geological history, why retain it to explain natural history in terms of separate special creations? Lyell having mapped out the intellectual territory to be traversed, the crossover from geology to biology appeared plain sailing to Darwin, whose way of thinking-by-analogy would seem to have been that, if there was a story of natural evolution in the geological record, so too might there be a similar story to tell in the study of sentient beings. As Darwin himself confessed, "The great merit of the *Principles [of Geology]* was that it altered the whole tone of one's mind and therefore that when seeing a thing never seen by Lyell, one yet saw it partially through his eyes."[13]

Darwin as Geologist

Darwin's frequent mentions of "the noble science of Geology" reflect-ed his long interest in and engagement with that subject, and Sandra Herbert has made the reasonable claim that "the primary reason why he is not well known as a geologist today is that the *Origin of Species*, read as a biological text, has outshone all else."[14] To many people, Darwin's earlier forays into geology might seem a trifle recondite, yet examination of this part of his career reveals important features of his working methods which continued to play a part in his biological work. In particular, the Darwin who could get carried away by certain metaphors was also capable of being misled by some inexact analogies. His first public forays into geological research provide examples of this tendency.

After his return from the *Beagle* expedition in the later 1830s, Darwin spent time putting together thoughts about evolution, which were to result in the provisional, unpublished pencil sketch of 1842 and in another, informal, private essay of 1844. Here were the seeds that would become the *Origin*.[15] In the meantime Darwin appeared in the public eye to be more active in the field of geological debate than in the biological sphere. Yet behind the scenes he was working on both fronts at the same time and there were many signs of intellectual cross-fertilization such that his geological work may be usefully compared with his methods in the biological sphere.

One of Darwin's early, pre-*Origin* forays into the sphere of public science concerned a geological puzzle in the Glen Roy area of the Highlands of Scotland, some twenty miles from Loch Ness, an area noted for the geological phenomenon of its three parallel "roads," as local folklore terms them. It is now known that the so-called roads, on a hillside in Glen Roy, are in reality loch terraces or strand lines that formed along the shorelines of an ancient ice-dammed loch at the time of the last Ice Age. The ice had repeatedly melted and refrozen over geological time with the water levels coming to rest at slightly different set points each time. In 1839, in a paper read before the Royal Society, Darwin unwisely chanced his arm by seeking to explain these roads as having resulted from ancient, *marine* beaches. But premier Harvard naturalist Louis Agassiz and Scots geologists soon showed that this must have been an ancient freshwater lake once dammed up by ice (the Swiss-American Agassiz had ample experience of glaciers in the country of his birth).

What is instructive about Darwin's swiftly disproved conjecture is that it was based on a misleading analogy he had come across in his voyaging years in South America. This he freely confessed in his autobiography when he wrote, "This paper was a great failure, and I am ashamed of it. Having been deeply impressed with what I had seen of elevation of the land in S. America, I attributed the parallel lines to the action of the sea; but I had to give up this view when Agassiz propounded his glacier-lake theory."[16]

As Robert Shedinger has observed, Darwin advanced his theory despite the telling absence of any ancient marine residues such as seashells, adding that "when Darwin developed what he felt was a compelling idea, he doggedly held to it even when faced with a lack of clear evidence."[17] This was a tendency readily observable in the biological sphere when he notoriously declined to recognize the true import of the absence of fossilized transitional forms as being detrimental to his theory of gradualistic evolution by natural selection with its claimed capacity to originate distinctly new species over a series of thousands and even millions of progressively differentiated generations, purportedly crossing a physiological barrier whose importance had been repeatedly underscored by such authorities as Cuvier in France and Richard Owen in Britain.

Argument from an inappropriate analogy was also to bedevil a second geological theory Darwin developed in 1842, this time in relation to the formation of coral reefs. During his travels in South America he had once observed what he took to be evidence that coral reefs emerged with the subsidence of surrounding land: as the land subsided, a coral reef or atoll would come to the fore. However, work by other geologists suggested that as often as not the reverse could be the case. That is, land underneath the sea would rise and bring towards the surface small organic forms congregating in reefs. In the end Darwin lost this debate as well. Crossing over to the biological sphere again, one is reminded of Darwin's wholly theoretical postulation of those hereditary entities he termed "gemmules," a theory which failed to find acceptance, since the postulation had no empirical support (as even Darwin conceded) and was definitively disproved by Mendelian genetics at the beginning of the twentieth century.

A Category Error

What of Lyell's resistance to Darwin's extending Lyell's uniformitarian, gradualistic, and materialistic methodology to biology? Although biological gradualism-cum-natural selection inspired by the idea of geological uniformitarianism seemed an uncontroversial form of intellectual progression to Darwin, Lyell thought that Darwin had carried over his early formation as a geologist into the biological realm indiscriminately and without attending to the appropriate modifications of analysis required. In Lyell's view, it was fundamentally misguided to conflate the organic and inorganic spheres in the way Darwin had, and this mistake doomed Darwin's theory of evolution to failure.

One can appreciate the force of Lyell's objection. One would not, for instance, think it appropriate to compare rocks and cliffs with human consciousness and view those entities as lying only slightly distant from each other on the same sliding scale. There is a great difference between planet Earth as a geological formation, which shares its history and mode of formation with the rest of the outer cosmos, and the later, superposed realm of terrestrial life and sentience. The one, after all, is quite simple when compared with the unsearchable complexities and subtleties of the organic world.[18] As Barry Gale once pointed out, "Mountains might decay and new mountains be thrust up again, but these new mountains were not considered to be more complex or very different from previous ones. For Lyell, there were no basic changes in the forms of natural phenomena."[19]

What Darwin proposed was qualitatively different, organic evolution occurring in creative patterns of progressive development—not chemicals mixing to form different compounds, or mountains fading into hills or plains, but amoebas becoming men. Since nothing of this sort was observable in Lyell's inorganic world of arbitrary forces churning away over untold ages, it is hardly surprising that Lyell thought the two domains incommensurable.

To some degree the dispute between Darwin and Lyell revolved around a matter of common sense. One did not need to be a renowned geologist to understand that like was not being compared with like by conflating geology and biology. Even the most basic proverbial wisdom balked at such an analogy, as is evidenced when one of Shakespeare's minor characters in *Julius Caesar* berates his peers for their lack of human feeling by calling out to them, "You blocks, you stones, you worse than senseless things." This accusation is presented by Shakespeare as an unarguable and absolute binary to distinguish insentient stones from human beings with their capacity for consciousness and fellow-feeling.[20]

Cosmologist Fred Hoyle described the same distinction in more scientific terms, noting that unlike the intricacy of the living world, in the case of inanimate objects, "there is no hierarchy of structure, with one level of subtlety piled on another. It is the existence of such a hierarchy which characterizes biological systems." His example: "A single crystal of rock, or of a mineral, or of a snowflake, contains subtleties of great interest. But such crystals do not fit together into larger patterns of still greater interest and complexity."[21]

Percival Davis and Dean Kenyon also touched on this issue when they observed that not all academic disciplines should necessarily be obliged to explain phenomena purely by reference to the same form of natural law. For such a fallacy "assumes a cookie-cutter view of science in which all disciplines ask similar questions and use the same 'scientific method.'"[22] In the case of the Darwin/Lyell *contretemps*, some analogies are clearly more apropos than others, and Lyell was convinced that he had to reject the applicability of the Darwinian analogy.

As far as we know, the living part of our planet has no identifiable counterpart in the external universe, *pace* unceasing attempts by space scientists to somehow conjure life from what appears to be the irredeemable barrenness of the extraterrestrial universe. Astronomers may yet discover an extrasolar planet with a complex biosphere, but that moment remains but a hope, and the more we learn about our galaxy and the wider universe, the more unusual our living planet appears, a trend that tends to confirm Lyell's view of the radical separation of organic and inorganic worlds. The sheer exceptionalism of our terrestrial biosphere stands in sharp contrast to the life-denying deadness of the outer cosmos. Indeed, even restricting our focus to planet Earth, the areas of our planet amenable to human habitation represent but a tiny fraction of the earth. As Michael Marshall noted, our atmosphere above a certain altitude will kill us (a fact well-known to elite mountain climbers) as would the heat of the crust only a tiny fraction of the way down toward the earth's core.[23] As for our place in the solar system, our "Goldilocks zone"—neither too hot nor too cold thanks to our distance from the sun—is all the more to be treasured for being such a very narrow band of habitability. Organic life is, in sum, exceptional in the extreme.

Our belated recognition of this fact has led to a considerable shift in many people's cosmological imagination. In no few cases it has resulted in a very sharp reversal of the once-ascendant cosmological worldview typical of philosopher Bertrand Russell's generation in the first half of the twentieth century. Russell (to whom Richard Dawkins acknowledges his philosophic debt) once famously described Earth as an accident in a cosmic backwater. More recently astronomer Carl Sagan dismissed it as a "pale blue dot."[24] But today, so complete is the discontinuity between Earth and the extraterrestrial dead zone revealed by modern space exploration, that it makes nonsense of the so-called "Copernican principle," according to which Earth is not the center of the universe, as it was believed in the pre-Copernican age, and is wholly unexceptional within the universe.

Nature's Destiny author Michael Denton has even gone so far as to suggest that the openly anthropocentric view held by our medieval forbears—that our world represented the geographic center of the universe—should now be rehabilitated, albeit under revision. To be sure, planet Earth is clearly *not* exceptional in being at the spatial center of the cosmos, but it certainly is so in the far more important symbolic and moral sense that we are the beneficiaries of a most un-usual planet on which *meaning* centers—a recognition that has proved little less than revolutionary in changing many hearts and minds.

Lyell of course knew nothing of modern-day cosmological re-searches, but the idea of Earth as a uniquely privileged planet was not unknown in his own day. William Whewell, an instructor and occasional peripatetic conversation partner of Darwin's when the latter was a student at Cambridge, had authored an influential volume in the early 1850s laying out reasons for the probability of Earth's being the only inhabited location in the universe.[25] Lyell then was hardly unique in failing to assent to Darwin's implied equivalence between geology and biology.

There is also this: When in 1872 Lyell declared the basic problem of creation/evolution to be inscrutable, he effectively joined hands with the later Wallace. Specifically, after the latter came to reject the idea that his own co-discovered theory of evolution by natural selection could provide a totalizing explanation for the full diversity of life and its many features, Lyell and Wallace—to Darwin's chagrin—became fast friends and intellectual allies.

Victorian Science's Manifest Destiny

But were Lyell and Wallace fighting a rear-guard action? Philosopher of science Stephen Dilley traces the appearance of methodological naturalism not to the Renaissance or the Enlightenment, but all the way back to Boethius of Dacia and his work *On the Eternity of the World*, "written during the height of the controversial assimilation of Aristotle's natural philosophy with Christian theology in the thirteenth century."[26] Dilley argues that Boethius argued for methodological naturalism not in some partial and nascent form, but methodological naturalism fully flowered.

Nor did it die with Boethius, only to be reborn centuries later. Although Boethius's views were condemned as part of Bishop of Paris Stephen Tempier's Condemnation of 1277, "within a century of the condemnation eminent scholars, like John Buridan and Nicole Oresme, were openly embracing methodological naturalism." Dilley continues, quoting Buridan and Oresme:

> "In natural philosophy," Buridan explained for instance, "we ought to accept actions and dependencies as if they always proceed in a natural way." In *Quodlibeta*, a work of natural philosophy, Oresme notes that "there is no reason to take recourse to the heavens, the last refuge of the weak, or demons, or to our glorious God as if He would produce these effects directly, more so than those effects whose causes we believe are well known to us."[27]

This methodological allegiance continued to gain steam in the Renaissance and early Enlightenment among natural philosophers and early scientists, from the specific arguments of such thinkers as Francis Bacon[28] as well as from the increasing pace of discoveries revealing natural laws and regularities where before was only mystery— as for instance in the work of the religiously devout Christian scientists Robert Boyle and Sir Isaac Newton.

After tracing the pedigree of methodological naturalism to earlier centuries, the distinguished historian of science Ronald Numbers summarized matters this way: "No single event marks the transition from godly natural philosophy to naturalistic modern science, but sometime between roughly the mid-eighteenth and mid-nineteenth centuries students of nature in one discipline after another reached the conclusion that, regardless of one's personal beliefs, supernatural explanations had no place in the practice of science."[29]

We should not, then, find it surprising that posterity has largely favored Darwin's view of things over that of Lyell and Wallace, and framed the matter as science extending its sway from one discipline to the next in a form of manifest destiny and by the same methodological means—the very transference which Lyell regarded as invalid. Thus, the currently ascendant view in origins science brushes Lyell's fine discrimination aside and sees the trajectory of scientific progress as crystal-clear: Newton having satisfactorily explained the starry heavens above, and Lyell having explained the realm of inanimate geology beneath our feet, the sights of scientific research were appropriately trained on organic life in quest of a solution to the riddles of biology in terms every bit as naturalistic as those offered by Newton in physics and Lyell in geology.

Under the aegis of that mighty scientific metanarrative, Darwinism was swiftly inducted into the post-Enlightenment domain labeled "science," and precious little in the way of cross-examination of the theory's fundamental tenets was thereafter permitted in academia. Hence by the later 1860s Lyell's uniformitarianism was extended and reformulated as biological gradualism, and as Richard Lewontin framed the matter, it came to be seen as a "completion of the unfin-ished Cartesian revolution that demanded a mechanical model for all living processes."[30]

As for Lyell's insistence on treating the animate and inanimate differently, it was summarily dismissed by those who positively *willed* there to be natural rather than supranatural answers to their questions about the world. Numbers comments:

> In his revolutionary essay *On the Origin of Species* (1859) Darwin aimed primarily "to overthrow the dogma of separate creations" and extend the domain of natural law throughout the organic world. He succeeded spectacularly—not because of his clever theory of natural selection (which few biologists thought sufficient to ac-count for evolution), nor because of the voluminous evidence of organic development that he presented, but because, as one Chris-tian reader bluntly put it, there was "literally nothing deserving the name of Science to put in its place." The American geologist William North Rice (1845–1928), an active Methodist, made much the same point. "The great strength of the Darwinian theory," he wrote in 1867, "lies in its coincidence with the general spirit and tendency of science. It is the aim of science to narrow the domain of the supernatural, by bringing all phenomena within the scope of natural laws and secondary causes."[31]

Hence when Darwin introduced his theory of evolution to the wider public, he was offering such persons the kind of answer they were eager for. Despite the lack of clinching empirical evidence for the theory, Darwinism was able to triumph in the decades after its intro-duction by synchronizing with the spirit of the age. In the context of the secularizing zeitgeist of the 1860s and beyond, it simply *felt* right.

Malthus and the Instrumentalization of Demography

Another figure that Darwin acknowledged his debt to is English cleric and economist Thomas Malthus. The following was Darwin's reac-tion to his reading Malthus's view that population growth generally outstrips the growth of food resources. This insight, combined with the idea of random variation in progeny, paved the way for Darwin's big breakthrough. As Darwin explained:

> In October 1838, that is, fifteen months after I had begun my systematic enquiry, I happened to read for amusement Malthus on Population, and being well prepared to appreciate the struggle for existence which everywhere goes on from long-continued observation of the habits of animals and plants, it at once struck me that under these circumstances favourable variations would tend to be preserved, and unfavourable ones to be destroyed. The result of this would be the formation of new species. Here, then, I had at last got a theory by which to work.[32]

Like a magpie building its nest, Darwin could see that he could, in a purely additive way, annex sundry thoughts to his present ensemble of ideas on evolution. Malthus meshed very satisfactorily with ideas Darwin had already picked up from his grandfather Erasmus, his geologist friend Lyell, and others: voilà!

Many of Darwin's contemporaries, however, found themselves bemused by Darwin's preoccupation with Malthus. Darwin's own son, Francis, himself to become a distinguished scientist, felt the whole matter of natural selection to be self-evident and expressed surprise that his father had found his reading of Malthus such a revelation when many others such as Erasmus Darwin, the Swiss naturalist Alfonsin de Candolle, William Paley, and Charles Lyell had already described the same struggle for existence in comparable terms.[33] Karl Marx's colleague Friedrich Engels too was unimpressed by the Darwin/Malthus connection, writing that it was not necessary to have consulted Malthus in order to perceive the struggle for existence since this was an idea which cohered intuitively with the politically uncharitable attitudes of the upper classes in nineteenth-century Europe. Engels viewed the Malthusian idea rather dimly as a tacit ratification of the individualist ideology of the arriviste merchants then acquiring a sphere of power in Europe.[34] In fact, Malthus's ideas hit such a chord of recognition that they were used as leverage to alter the older but more generous Elizabethan Poor Law in Great Britain. (Under the new law the poor had to compete for work or else be consigned to the workhouse.[35])

But what such commentators as Darwin's son, Friedrich Engels, and others perhaps did not fully realize was that Darwin had a self-interested motive to appropriate Malthus by extrapolating from his writings the germ of an idea supportive both of his grandfather's evolutionary ideas and of his own theory of natural selection. What interested Darwin, who had already lobbied Lyell for his prestigious support (unsuccessfully), was in part the chance to co-opt the respected demographer in order to buttress and give prestige to his own ideas. This was, at least in part, a case of name-dropping.

What is remarkable about the block quote above is Darwin's revelation that his reading "for amusement Malthus on Population" is what triggered his eureka moment for his theory of evolution by natural selection. Needless to say, Malthus had not advanced or even hinted at the obviously ludicrous idea that economically successful members of human societies might somehow morph over time into transhumanist *Übermenschen*. Such ideas of human transfiguration would have been far from his thoughts as an economist. That particular idea was clearly Darwin's creative application of Malthus's insight, not anything Malthus himself endorsed.

Darwin's elliptical reasoning for the awkward conjunction only makes sense in the context of more than a century of transmutational speculation he was privy to via his grandfather and of which he regarded himself as the family legatee and trustee. Darwin, in other words, was eager to enlist Malthus as an authority buttressing the conjectures of himself and his grandfather.

Of course, there was no objective need to invoke the name of Malthus in discussions of evolution. Many had debated similar issues without having recourse to the noted demographer. But a curious aspect of the publication of the *Origin* was that Darwin, preparing himself mentally for imputations of impiety, was instead caught off-guard by accusations of plagiarism. After 1859 a number of persons came forward to stake a claim to having discovered the very phenomenon which Darwin had regarded as his own intellectual fiefdom, natural selection.[36] Rebecca Stott has even gone so far as to conclude that Darwin's theory "was just as much the discovery of British clergymen, doctors, fruit-farmers and gentlemen naturalists working away with microscopes in the British provinces."[37] One of the "contenders" for Darwin's crown, the prosperous Scottish horticulturalist Patrick Matthew, even made the express declaration that the conception of natural selection came to him "intuitively as a self-evident fact, almost without an effort of concentrated thought."[38] He certainly did not need Malthusian spectacles to consider the matter.

Chronologically first among the modern claimants was William Wells, who in 1813 read out a paper to the Royal Society of London touching on the subject of African ancestry having the advantage of conferring immunity to certain diseases. Some would bear disease better than others, he said. Those would consequently multiply, the others decrease, not only from their inability to combat the disease but also from their disadvantage in contending with their stronger peers.

The aforementioned Patrick Matthew put forward his survival-of-the-fittest idea in an appendix to a somewhat recondite publication called *Timber and Naval Arboriculture* in 1831. I give here a relevant extract from that section of the book:

> As Nature, in all her modifications of life, has a power of increase far beyond what is needed to supply the place of what falls by Time's decay, those individuals who possess not the requisite strength, swiftness, hardihood, or cunning, fall prematurely without reproducing—either a prey to their natural devourers, or sinking under disease, generally induced by want of nourishment, their place being occupied by the more perfect of their own kind, who are pressing on the means of subsistence.[39]

Four years later the young naturalist Edward Blyth wrote in the following terms about artificial selection:

> When two animals are matched together, each remarkable for a certain peculiarity, no matter how trivial, there is also a decided tendency in nature for that peculiarity to *increase*; and if the produce of these animals be set apart, and only those in which the same peculiarity is apparent, be selected to breed from, the next generation will possess it in a still *more* remarkable degree; and so on, till

at length the variety I designate a *breed*, is formed, which may be very unlike the original type.[40]

Two years later, making a similar point, he wrote, "May not, then, a large proportion of what are considered species have descended from a common parentage?"[41]

But shortly thereafter Blyth felt impelled to retreat from his own conjecture. He realized that in those circumstances living species would blend into each other, a form of hybridization not observable in nature.[42] As he put it in the later essay:

> Were this self-adapting system to prevail to any extent, we should in vain seek for those constant and invariable distinctions which are found to obtain. Instead of a species becoming gradually less numerous where its haunts grade imperceptibly away, we should discover a corresponding gradation in its adaptations; and, as the most dissimilar varieties of one species (those of the dog, for instance) propagate as readily together as individuals of the same variety, producing offspring of blended characters, so much so, that human interference is requisite to preserve a breed unadulterated, the unbending permanency of the distinguishing characteristics of all wild animals becomes of double import.[43]

But what about an animal form that has spread into widely diverse biological niches? Might not it evolve into radically new forms? Blyth noted that, besides the fact of our being unable to point to a series of actual creatures grading imperceptibly from one form to another, there is also positive evidence against the idea—namely, examples of creatures that vary only slightly despite being dispersed into widely different ecological niches. "The common jay is diffused over a wide range of latitude, but is the same in Italy as in Sweden: this would not be were it affected by locality or climate," he wrote. And when he says "were it affected," the context suggests he means greatly transformed over time.

Blyth continues: "The very trivial distinctions, therefore, which characterize it apart from that of Japan, and from that of Asia Minor, we are warranted in esteeming of specific value."[44] Of specific value, that is, as a strong indicator of the strict limitations of evolution by natural selection.

In all this Blyth is in the intellectual tradition of the great eighteenth-century French naturalist Georges Cuvier, who concluded that animal life could *not* be fitted into a unilineal ascending system. Diverse animals constituted a bush rather than a ladder, since they belonged to distinct groups: vertebrates, mollusks, articulates, radiata, et al. The mollusk body plan could never "transition" into the vertebrate one, because the differences between the two types were insuperable. Why? Because the main set of features that made a biological form distinct from any other biological form were interdependent. Cuvier elaborated thus:

> For example, if an animal's teeth are such as they must be, in order for it to nourish itself with flesh, we can be sure without further examination that the whole system of its digestive organs is appropriate for that kind of food; and that its whole skeleton and locomotive organs, and even its sense organs, are arranged in such a way as to make it skillful at pursuing and catching its prey. For these relations are the necessary conditions of existence of the animal; if things were not so, it would not be able to subsist.[45]

This interdependency of parts, which came to be known as Cuvier's Law of Correlation, suggests that biological form could not be altered gradually and piecemeal, at least not indefinitely. Too many interdependent parts need to be altered simultaneously to ensure the viability of the creature. In short, the barrier to unlimited naturalistic evolutionary change was a real one.

Blyth, having identified the mechanism of natural selection and considered the possibility of unlimited change along these lines, corrected himself by such considerations—in other words, by the kind of reality check which Charles Darwin chose to override.

It is not as if Darwin were unaware of such difficulties facing his theory. In an admirably candid passage, Darwin even tacks on the related problem of the lack of fossil evidence, which further compounds the difficulty of accepting his ideas: "Why, if species have descended from other species by insensibly fine gradations, do we not everywhere see innumerable transitional forms? Why is not all nature in confusion, instead of the species being, as we see them, well defined?... as by this theory innumerable transitional forms must have existed, why do we not find them embedded in countless numbers in the crust of the Earth?"[46]

"Why indeed!" one might reply. Darwin, in answering his own question, appealed to the incompleteness of the fossil record. Darwin held out hope that future fossil discoveries would mitigate the problem of missing transitionals. If anything, however, the problem has grown worse as paleontologists have developed a more complete picture of ancient life and fossil succession.[47] This development is a further example of a stubborn disparity between Darwin's theory and the uncooperative facts of objective reality.

The Legacy of Antiquity

An irony in Charles's efforts to fortify the Darwin brand by aligning it with the supposedly bang-up-to-date foundation of Malthusian economics is that it had already been anticipated more than two thousand years ago, in the time of ancient Greece and Rome.

Cambridge professor Tim Whitmarsh recently complained that our modern ignorance of the classical heritage has allowed a "modernist mythology" to take hold, according to which eighteenth- and nineteenth-century Europeans were the first to ponder evolutionary matters or "do battle" with the gods.[48] Whitmarsh was not the first to argue otherwise.

Shortly after publication of the *Origin* in 1859, many among the classically educated British elite viewed Darwin's purportedly new evolutionary ideas with some skeptical bemusement for being a trifle passé. "I cannot understand why you scientific people make such a fuss about Darwin. Why, it's all in Lucretius!," yawned premier Victorian educator and poet Matthew Arnold to a biology professor in his circle.[49] Surely anybody with knowledge of the classical world of thought must know, Arnold protested, that the Greek Epicurus and his Roman follower, Lucretius, taught that the answer to the world's awe-inducing complexity was to be sought not in a once-and-for-all divine creation but in different shapes and objects generated at random by the chance interaction of atoms (such in brief being the ancient philosophy of atomism). Plants and animals had simply "evolved" via an extended process of trial and error. Thus, many centuries before 1859 a largely random process of evolution had been posited as being responsible for the slow evolution of all Earth's sentient species. In some cases that evolutionary journey had been unsuccessful, went the ancient contention, resulting in creatures not properly equipped to compete for resources or to produce offspring. Such creatures were destined to extinction, in contradistinction to vigorous and perfectly formed specimens able to adapt and reproduce.[50]

Such philosophical speculation going back more than two millennia had effectively given the modern world a simple either/or choice. Darwin in effect chose the minority Epicurean/Lucretian model whose most celebrated British proponent was the Scottish philosopher David Hume.[51] It was Darwin's favoring of the Lucretian explanatory model—a model that Darwin discreetly and perhaps grudgingly buttressed with a deistic underpinning at times—which came to erode the Aristotelian/Christian conception of ultimate reality, at least for large sections of the intellectual classes. The atomistic conception that Aristotle, Plato, Cicero, the early physician Galen, and a host of further distinguished thinkers had denounced for what they regarded as its manifest absurdity was now, phoenix-like, rising from the ashes to which it had been relegated by the leading thinkers of the early Western tradition for well over a thousand years. This factor is foundational to an understanding of Darwinism, and a brief account of those ancient philosophical voices is necessary to clarify the matter.

Epicurus and Lucretius

For Epicurus and his fellow atomists, nature amounted to little more than a series of chance combinations and recombinations—that fuzzy and ill-understood process even today nebulously described as "complexification." In short, for the ancient atomists, life was not fundamental or aboriginal to the world, nor had it been created by any purposive intellect, but instead developed as emergent properties of particular types of atomic configurations. Such notions clearly anticipate Darwin's idea of life originating in a "warm little pond" via fortuitous aggregation of chemicals and energy followed by biological evolution via random variations and survival of the fittest. The modern culture wars supposedly unleashed for the first time by Darwin had, in point of historical fact, had an ancient counterpart in the Greek atomists, whose views were vigorously debated in the classical world. Materialist philosophers in antiquity pitted themselves against the teleological convictions of such leading philosophers as Plato and Aristotle, who argued for a creative mind (*nous*) and an original Unmoved Mover, the conception later endorsed by Thomas Aquinas and incorporated within Christian theology in the Middle Ages.

Strange as it might at first seem, Darwinism, when viewed from a purely philosophical perspective, might best be understood (as Arnold suspected) as a late sub-branch of ancient speculative thought rather than as science in the modern sense, with its demands for rigorous demonstration rather than unsubstantiated conjectural flights. To be sure, Darwin was attuned to the received modes of scientific reasoning of his day, including those championed by John Herschel, William Whewell, and (earlier) Newton; and he worked mightily to provide concrete evidence for his theory. But a close reading of his intellectual journey suggests that it was his speculative flight, rather than any empirical investigation, that (to use a mixed metaphor!) wagged, and carried airborne, the proverbial dog.

It should be stressed, however, that there is no evidence that Darwin ever read the ancient atomists. Although he could read Greek and Latin, and had written one of his three Cambridge exam papers in Classics, he would not necessarily have read and studied the works of those authors. It is quite likely, however, that he had encountered their basic ideas via general reading and conversation, and perhaps also via his grandfather, the scholarly Erasmus Darwin, with his much superior knowledge of the classics.

If Charles did encounter those ideas as a young adult, either directly or indirectly, "he might have recognized..." in the words of Rebecca Stott, "in corners of Epicurus, Democritus, and Empedocles the glimmerings of thoughts and questionings that were remarkably similar to his own."[52]

Ancient and Modern Culture Wars

As observed above, the overwhelming majority of philosophers and theologians of both the classical and Christian centuries went back in their intellectual origins to Plato, Aristotle, and the Stoics, and to one version or another of the idea of a divine artificer. The pioneering medical authority of the ancient world, Galen, had little but derision for the atomists. For him, those philosophers' unqualified exclusion of divine causation left nature stranded and unintelligible. Even though members of the atomist school were wont to appeal to the *vis infinitatis* (the power of infinity or a limitless time/space-scale), they were still obliged to postulate accident on a risibly vast scale, objected Galen.[53]

Cicero had made the same point as the pioneering physician:

> I cannot but express astonishment at this, that anyone could convince himself that certain solid, indivisible bodies are borne along by their thrust and weight, and that from the chance collision of these bodies is created a universe supremely embellished and beautiful. In my view, anyone who imagines that this could have happened, must logically believe that if countless numbers of the twenty-one letters of the alphabet, fashioned in gold or in some other substance, were thrown into the same receptacle and then shaken out upon the ground, they might form the *Annals* of Ennius [a now lost Roman history] made immediately readable before our eyes. Yet I doubt if as much as a single line could be so assembled by chance.[54]

Fred Hoyle, a one-time atheist, updated the argument more than 2,000 years later:

> A generation or more ago a profound disservice was done to popular thought by the notion that a horde of monkeys thumping away on typewriters could eventually arrive at the plays of Shakespeare. This idea is wrong, so wrong that one has to wonder how it came to be broadcast so widely. The answer I think is that scientists wanted to believe that anything at all, even the origin of life, could happen by chance, if only chance operated on a big enough scale. This is the obvious error, for the whole Universe observed by astronomers would not be remotely large enough to hold the horde of monkeys

needed to write even one scene from one Shakespeare play, or to hold their typewriters, and certainly not the wastepaper baskets needed for throwing out the volumes of rubbish which the monkeys would type. The striking point is that the only practicable way for the Universe to produce the plays of Shakespeare was through the existence of life producing Shakespeare himself.

Despite this, the entire structure of orthodox biology still holds that life arose at random. Yet as biochemists discover more and more about the awesome complexity of life, it is apparent that the chances of it originating by accident are so minute that they can be completely ruled out. Life cannot have arisen by chance.[55]

Moving forward from Cicero and Galen, a little more than a millennium later Thomas Aquinas argued that there was a universal consensus on the grounds of logic alone that there must be an intelligent source of all things. That is, he insisted that it was by no means necessary to be an adherent of the Judeo-Christian tradition to accept the idea.[56] The understanding became a leitmotif of Western thought in ensuing centuries, such that for instance in the philosophical writings of Lord Herbert of Cherbury in the seventeenth century, the logical necessity for an initiator of all things was characterized as a "truth of first inscription."[57]

The opposing, Lucretian attempt to explain reality on an antiteleological basis, together with the related claim that the world only *looked like* it had been crafted by some designer, would be mediated to the modern world by Scottish philosopher David Hume. But just as Lucretianism remained a minority view in antiquity, so Hume's view failed to gain much traction either in the late eighteenth century or in the first half of the nineteenth. In fact, Lucretianism did not begin to be taken seriously as a respectable postulate until Darwin came onto the scene to give it a form of cover. The world after the Enlightenment needed something with at least the appearance of scientific rigor to validate the meditations of philosophers. Enter Darwin's theory of evolution, stage left.

Lucretianism and Darwinism, the ancient and modern hypotheses on their own might have appeared tenuous, but when conjoined their resultant force had the startling effect of being able to overturn older, tried-and-tested ways of thinking. To those unsympathetic to Darwin it must have appeared like a perverse case of two wrongs coming together to make a right.

Atomism as instrumentalized by Epicurus and his successors was, as David Sedley remarked, employed as a "weapon against divine creation."[58] If all operative power in the world was down to chance, ran the argument, then no gods would be able to harness any power of their own to oppress us. In one respect one can understand and even sympathize with the atomists' argument from a purely tactical point of view. From all that we know from Homer and later Greek and Roman mythologists, the gods and goddesses of popular conception were hardly more than fallible human beings writ large. Such figures, devoid of moral sublimity, had become a source of embarrassment to thoughtful Greeks and Romans. Lucretius contended that they inspired fear rather than allegiance and were more to be propitiated than venerated. Such, for instance, was the infamous case in Homeric tradition of Agamemnon, whose fear of the goddess Artemis caused him to sacrifice his own daughter at the goddess's whim in order to secure a favorable wind for his fleet's sea voyage to Troy.

Epicurus was an atheist in the original sense of the word, an anti-theist, one who rejected the baleful and destructive values of the Athenian pantheon. His was more a declaration of war against the flawed moral nature of the gods (technically termed "theomachy" or "misotheism") than it was a statement of outright disbelief (a-theism).[59] He was often in fact referred to by his contemporaries as "Epicurus *theomakhos*,"[60] battler of the gods.

Today, thankfully, we have come a long way from the times of child sacrifice and other cruel propitiatory rites to pagan gods, yet the battle persists.

Some two decades ago, atheist Richard Dawkins paid to have the somewhat underwhelming injunction "There's probably no God, now stop worrying and enjoy your life" emblazoned on the side of London buses. Those who held to considerably more benign conceptions of the deity undoubtedly were bemused and prompted to ask themselves what precisely they might have to be "worried" about.[61] It was as if we were all living in the time of Epicurus, where conceptions of the gods as capricious, amoral, and even hostile to humankind were commonly held. Perhaps Dawkins had in mind biblical hellfire and damnation, a theological issue that seemed to have loomed large with Darwin himself. Be that as it may, the sentiment emblazoned on the bus, in conjunction with Darwin's divinity-defeater, natural selection, appears to have had a most potent effect on Western culture, and long before Dawkins's cheeky marketing campaign involving London's mass transit system. Indeed, thanks to Darwin, the atomists, "with their faltering anticipations of Darwinism," writes Sedley, "may for the majority of readers have emerged as today's winners by proxy."[62]

Darwin, although largely innocent of formal philosophy himself, was able to pull off a dazzling intellectual coup against the majority of thinkers of repute in the Western tradition going back over two millennia. The coup is all the more surprising since Darwin succeeded where all the argumentative brilliance of David Hume had struck little fire. How so? There were undoubtedly a number of reasons why Darwinism was able to gain influence, some particular, some general. Let us begin with the general one of the shifting philosophic and religious landscape of Britain between the 1820s and the 1860s. This shift in temper occurred at various levels of British life, being observable at the elite level, the grassroots level, and in that genre of creative writing which might be termed the literature of doubt, to which I turn first.

4. Victorian Culture Wars

A USEFUL MEMOIR ON THE EARLY RECEPTION OF DARWIN'S *ORIGIN* was left us by the acclaimed author who by both birth and marriage was plugged into the nineteenth-century zeitgeist like few others, Mrs. Humphry Ward (born into intellectual aristocracy as Mary Augusta Arnold), the author of a particularly moving novel about loss of faith, *Robert Elsmere* (1888). In looking back at her experiences of Oxford in the 1860s/1870s, Ward noted that "the men of science entered but little into the struggle of ideas that was going on. The main Darwinian battle had been won long before 1870; science was quietly verifying and exploring along the new lines; it was in literature, history and theology that evolutionary conceptions were most visibly and dramatically at work."[1]

Her estimation that the main battle over Darwinism had been won by 1870 is certainly debatable, but it's surely correct that the tide by that point had turned in Darwin's favor. And Ward is surely correct to expand our perspective as we consider the intellectual milieu in which Darwinism flowered. It is to sources beyond the scientific—literature, history, theology—to which we must go to fully understand how Darwinism won the hearts and minds not just of Victorian scientists but of the late Victorian world more generally, triumphing over the scientific objections exhaustively spelled out by the early expert reviewers.

First, from the perspective of the cultural producers and commentators identified by Ward, Darwinism will have worn a rather different aspect to that observed from the unblinking perspective of empirical science. Within that larger philosophic context there had emerged over several centuries a succession of voices calling for God's dethronement, beginning with Baruch Spinoza in the seventeenth century, proceeding via Edward Gibbon, David Hume, and Jean-Jacques Rousseau in the eighteenth century and thence through to Ludwig Feuerbach, Arthur Schopenhauer, and others in the nineteenth century, foremost of whom in England was the poet Algernon Charles Swinburne.[2] The cumulative nature of that developing narrative meant that, by the middle of the nineteenth century, educated society may have been more open to Darwin's idea than it would have been in previous generations.

Certainly it's difficult to ignore the evidence that many Victorians very much wished Darwin's theory to be true. On the slightly dubious principle that empirical facts should never be allowed to get in the way of a good story, many turned a blind eye to the scientific inadequacies revealed by the expert reviewers, and the theory was glossed as the all-but-inevitable extension of a long-nourished intellectual project. The Darwinian narrative was co-opted and integrated into that wider philosophic argument so as to lend it the prestigious imprimatur of *science*.

So, if we have a dog wagging a tail here, who is the dog and what is the tail? It is unlikely that the atheistic narrative built up by generations of philosophic voices was alone adequate to give the scientifically excoriated theory of Darwinism the pass it came to receive. As Alec Ryrie aptly pointed out in *Unbelievers: An Emotional History of Doubt*, "Intellectuals and philosophers may think they make the weather, but they are more often driven by it."[3]

The more decisive forces in the eventual acceptance of Darwinism may instead have issued from works of imaginative literature with a wider outreach than that achievable by formal philosophy. Doubt arose in popular sentiment long before it was translated into formal philosophical terms, Ryrie suggests, its emotional contours being perfectly visible to most before it was endowed with precise conceptual shape in the high-culture discipline of formal philosophy.

Literary Theomachy

"A novel can present ideas in a way more dramatic, engaging, and hence threatening than countless nonfictional volumes of political philosophy," commented philosopher of science Michael Ruse.[4] In place of "novel" he could aptly have spoken of literature generally. And one could add to that observation the point that many an author has found himself willing to express heterodox opinions in the context of imaginative literature that he would never dare have expressed in straightforward philosophical or theological argument.

Thus do we find that in matters of metaphysical rebellion, as Bernard Schweizer observed, the imaginative space of literature has down through the ages provided "the principal conduit for expressions of animosity towards the Almighty."[5] We find such veiled expressions of protest to include the early thirteenth-century German *Poor Henry (Der Arme Heinrich)*, a short verse novella with strong echoes of the Book of Job. When Heinrich, the exemplary lord of the manor, is unaccountably stricken with leprosy, his feudal dependents express their frustration that they are unable to curse the person responsible for their beloved lord's suffering, as it had undeservedly been visited upon him by God.

In the same tradition of coded protest theology stands the early fifteenth-century *Death and the Ploughman (Der Ackermann aus Böhmen)*, an imagined courtroom dialogue of the eponymous plaintiff with the accused, Death, over the premature loss of the ploughman's wife. The ultimate target of the ploughman's grief-stricken wrath is of course readily apparent. Similar examples might be adduced from the older literature of many other cultures, but of more immediate relevance to Darwin's generation were those writers who can be traced in a fairly direct line from William Blake and Percy Bysshe Shelley through Algernon Charles Swinburne, James Thomson and, near the nineteenth century's end, Thomas Hardy.

Consideration of relative chronology makes it clear that the publication in late 1859 of Darwin's *Origin* did not *cause* the Victorian crisis of faith but rather served to confirm the skepticism of those who had already formed anti-theistic attitudes on other grounds. Swinburne (1837–1909) provides an approximate chronological fit with Darwin. Swinburne is not read by many today, and many people may know little of him other than as a melodious versifier with some vague connection with the pre-Raphaelite group of artists. This scarcely does justice to the iconoclastic stature of "the laureate of mid nineteenth-century unbelief"[6] who in his verse drama *Atalanta in Calydon* has his classic chorus protest against "the supreme evil, God."

Swinburne, an admirer of Epicurus, is thought to have lost his faith while studying in Oxford in 1858/9 (just before the publication of *Origin* on November 24 of 1859).[7] He shared Epicurus's disdain for the old Greek pantheon, seeing no good reason why humans should worship deities (pagan or Christian) who had not shown themselves worthy of their supposed creatures. Instead, he went back imaginatively in his pagan turn to ancient personifications of natural forces such as Proserpine, goddess of the seasons and natural cycles. Meanwhile, in his *Laus Veneris* (Praise of Venus) he reprises the medieval Tannhäuser legend in which his affirmation of the processes of sex and procreation implies criticism of the sexual puritanism of contemporary denominations of the Christian Church. Against what he saw as the morbidities of Christian asceticism (he famously describes the Christ figure as that "pale Galilean" turning the world grey with his breath),[8] he foregrounds the robust life force represented by the ancient goddess of love.

Bernard Schweizer even sees Swinburne as a precursor of Friedrich Nietzsche, since both "saw in Christianity a religion in decline and both advocated the extermination of God in order to infuse new vitality into the spiritual life of European culture."[9] Perhaps so, but what is certainly true is that, unlike Ward's figure of the doubting clergyman, Robert Elsmere, who internalizes his problems with the Almighty by throwing himself into good works in the impoverished East End of London, Swinburne, exulting in his role of *enfant terrible* to mid-Victorian England, externalizes his disaffection by directing his ire squarely on Him whom he took to be its proper target.[10]

The Swinging 1860s

In some respects the 1860s resemble the 1960s with the long-haired rabble-rouser Swinburne reminiscent of some pop-stars of the Flower Power era.[11] (Some Oxford undergraduates of the time were observed linking arms and declaiming parts of Swinburne's heterodox poems as if in a Greek chorus.[12]) A growing number of persons in the mid-nineteenth century simply wished to close their account with the divine. To this end many were naturally eager to try to understand the world via exclusively secular criteria. Among Swinburne's like-minded coevals, we might mention Auguste Comte, whose secular-ized "religion of humanity" had a pan-European influence (not least in England), Karl Marx and Friedrich Engels, the novelist George Eliot and, in the later nineteenth century, the novelist Thomas Hardy.

The setting aside of a First Cause may help to explain why a theory at first adjudged scientifically unviable was nevertheless able to achieve the status of unquestionable shibboleth it later acquired and still enjoys to this day. Viewed in its chronological context, the *Origin* belongs to the same genre philosophically as those other nineteenth-century works of literature discussed by A. N. Wilson, which cumulatively announced "God's funeral."

Without that "supporting cast" of literary works and the frame of mind they both reflected and encouraged, it is difficult to imagine the *Origin* gaining the uncritical acclaim it went on to receive. Many who embraced the accumulated improbabilities of Darwinism did so only after losing faith in religious explanations on other grounds. People were beginning to look for some rallying point in the closed world of science to explain the wider world in a way that satisfied the secular-izing impulse. If this left out of account any final-cause thinking and so offended the strict logic of cause and effect, so be it. Many were in effect prepared to "cut off their nose to spite their face."

Darwin himself exemplified what may be loosely termed this Argument from Pique to a tee when he once wrote that he could not, on humanitarian grounds, see why anybody would even *wish* Chris-tianity to be true.[13] As we saw in Chapter 2, in a letter to American academic Asa Gray, Darwin doubled down on this thought, writing,

"I cannot persuade myself that a beneficent and omnipresent God would have designedly created the Ichneumonidae with the express intention of their feeding within the living bodies of caterpillars, or that a cat should play with mice. Not believing this, I see no necessity in the belief that the eye was expressly designed."[14]

In this somewhat muddled mixture of biological and theological thinking,[15] Darwin sought to convince himself on grounds of moral sensibility that God could not have been an active shaper of evolution. It need hardly be pointed out that this conclusion is not a logical inference but more a response to that growing groundswell of religious disaffection in mid-Victorian Britain in which Darwin grew to maturity.

The shaky logical basis of Darwin's thinking has not gone entirely unremarked. The notion of a supposedly unintelligent yet remarkably independent and self-directing biosphere (like the postulation of a self-creating cosmos) presents, when dispassionately considered, an offense to logic great enough to invite attempts from the sympathetic to square the circle. Simon Powell makes such an attempt in *Darwin's Unfinished Business: The Self-Organizing Intelligence of Nature*. Powell willingly concedes that "to state nonchalantly that evolution just happens and that it involves no more than changes in a gene pool over time, or that it is simply descent with modification, is really not good enough. Nature is crying out for a more decent appraisal."[16]

Indeed so, yet Powell's move of attributing what he terms "bio-logic" to nature, now declared by him to be *intelligent*, can hardly be said to advance a fresh *naturalistic* explanation. For the contention begs the question of the *origin* of such intelligence. Unless one subscribes to Spinoza's conflation of God with Nature (*Deus sive Natura*) it is difficult to accept Powell's theory. He explains elsewhere in the volume that his thinking owes much to Christian de Duve's postulation of a "cosmic imperative." Yet de Duve himself conspicuously failed to overcome the logical obstacle of explaining how an imperative could issue *itself*. We find ourselves yet again in the realm of a "suppression of the agent" and hence of crypto-magical automatism.

Winning Over the Intellectuals: Oxford 1860

In that conservative bastion of British life, the University of Oxford, changing times were tolerated only under protest and belatedly. Although Oxford as an institution played no role in the formation of Darwin's evolutionary ideas, it became the epicenter of the storm which was to surround his theories.

On June 30, 1860, under the auspices of the British Association, the just-opened New Museum of that venerable university became the site of a legendary debate between Thomas Huxley (deputized for Darwin) and Bishop Samuel Wilberforce. Darwin was not present at the meeting. He had been suffering with acute gastro-intestinal problems for some two decades and conceded he was not so good a public man as the forceful Huxley. Darwin, given his habitual modesty, might well have concurred with William Irvine's observation in his joint biography of Darwin and Huxley that, where Huxley could hold an audience with conviction and eloquence, Darwin's public utterances were more apt to "terminate in unintelligibility and syntactic disorder."[17]

The precise details of the Oxford meeting have become shrouded in a degree of folklore, but we do know that, after a "warm-up act" in the shape of a general lecture on Darwinian ideas by John William Draper of the United States, some members of the audience rose to respond. Among their number was none other than (now) Admiral Robert FitzRoy, Darwin's erstwhile skipper on the globe-circling *Beagle* expedition in the 1830s. FitzRoy, clearly not enamored of his former sea companion's blasphemous views, proceeded to denounce Darwin somewhat incontinently for having been a "viper" on board his craft more than a quarter of a century previously.

At some point in the excitement a Lady Brewster is recorded as having fainted (although this was high summer *sans* air-conditioning); and the disputatious affair is reputed to have reached an inglorious height when Wilberforce demanded to know of Huxley whether Huxley were descended from an ape on the side of his grandmother or grandfather. Huxley's on-the-spot response (which Darwin could not have dreamed up in a month of Sundays) was to the effect that he would rather have an ape for a grandfather than a man who abused his higher intellectual faculties to introduce ridicule into a serious scientific discussion. Opinions differ on which side came out on top in the debate; the legend that Huxley's improvised riposte delivered the coup de grâce appears unlikely. What all seemed to be agreed on was, however, that, histrionic asides and strident noises apart, Huxley and the Darwinian side were accorded a fair hearing. Such fair-mindedness seems entirely unremarkable by present-day standards, yet against the backdrop of earlier Victorian Oxford it was significant and requires a little historical contextualization.

On the face of it, the entrenched conservatism of Oxford, with its staunch Anglican traditions and quasi-monastic colleges peopled exclusively with unmarried dons, seemed an unlikely venue for the airing of "advanced" and challenging views in matters touching religious allegiance. In the first half of that century the great majority of college fellowships could be held only on condition of the celibacy of the incumbent. A don in this period was by definition a clergyman since holy orders were a precondition of his holding a fellowship. Typically, the system was that most fellows were elected shortly after taking their B.A.s, and they would then take their Orders whilst awaiting a college living, many holding curacies elsewhere in advance of their formal college appointments. Hence the accredited ideal of life and thought for Oxford's academics in the first half of the nineteenth century represented a fusion of Church of England orthodoxy with that more general expectation of Victorian high-mindedness.

The roots of this ethos ran deep. Already by the early modern period the colleges had evolved into bastions of Anglicanism conservatively interpreted—so conservatively, in fact, that from 1833 to 1845 Oxford was riven by a conflict which, to judge from many of its protagonists' favorable attitudes to the Catholic Church, even threatened to put the clock back centuries by transferring the seat of religious authority from Canterbury to Rome, that is, to the status quo before the Protestant Reformation.

In the 1830s a group of reforming Anglican academics and clergymen, John Henry Newman, John Keble, Edward Pusey, and their sympathizers, were growing increasingly worried about the establishment's lack of seriousness regarding its religious duties. The Tractarians (also known collectively as the Oxford Movement) attacked what they saw as the doctrinal laxity and disregard for the church's ancient heritage on the part of many congregants. Richard Hurrell Froude, a young associate of Newman and Keble, objected to the incorporation of the church within the state as its official religion. That rapprochement, thought by many to be the church's ultimate source of strength, was in Froude's view tending to undermine the church's ancient doctrinal purity. This, it was charged, had led to the present stance of Erastianism, that is, the willingness to subordinate the claims and prerogatives of the church to the requirements of state policy.

The suspicion inevitably arose that the high church Tractarians were stalking horses for Roman Catholicism, a suspicion strengthened when Newman published a tract arguing that the 39 Articles, the foundation of belief and observance for the Anglican church, were, rightly understood, compatible with the Roman church. Suspicions of a "Popish plot" then appeared confirmed when Newman converted to Roman Catholicism in 1845.

To modern eyes the Tractarian movement may appear to be an absurd fit of medievalish navel gazing, but that is to view the situation in the unsympathetic light of secularizing tendencies which influenced Oxford only *after* mid-century. The confessional fissures it opened were real enough at the time and had real consequences, even causing major schisms in the Wilberforce family between those siblings sympathetic to Rome and those (like Samuel himself, the future Anglican Bishop of Oxford) who continued their allegiance to Canterbury. Hence, indulging in an historical hypothetical, we may speculate on what might have been the effect had Darwin, instead of sitting on his magnum opus for more than two decades, published the *Origin* when some of his friends and colleagues thought he should have, in the 1840s not long after his first pencil sketch of the *Origin*.

We know that the ultra-conservative religious ethos of Oxford in the earlier decades of the nineteenth century was not welcoming of the German Higher Criticism of the Bible, which submitted sacred texts to the same kind of forensic evaluation as it did secular ones, accepting a thoroughly naturalized view of the biblical text as a starting point for inquiry. The forensic modes of analysis used by classical scholars to study Thucydides or Herodotus were not thought by Oxford incumbents to be suitable for application to biblical texts, which remained very much "ring-fenced" in the minds of the Anglican dons of the first half of the century. With that analogy in mind, it seems unlikely that Darwin's work would have gained the resonance it was to achieve in 1859/60 had he rushed into print in in the 1840s. The more favorable hearing it achieved may in good part be attributed to the change of intellectual temper occurring in the relatively short space of time when Oxford, in tandem with many other sections of the educated world, had moved on from the defensive ecclesiastical positions it had taken up in the earlier decades of the century.

From about the middle of the nineteenth century both Oxford and Cambridge began the laborious and contested[18] process of acknowledging the advent of modernity by transforming themselves from their previous status as essentially clerical institutions. Until 1854 the constituent colleges of Oxford University had been governed by statutes put forward by Archbishop William Laud in the early seventeenth century, and upon matriculation students were required to subscribe to the 39 Articles of the Church of England. By the 1870s, the Anglican monopoly of Oxford and Cambridge was loosening its grip, and the self-evident link between academia and clergy was less firm. In 1871 Prime Minister William Gladstone was instrumental in repealing the much-resented Test Acts (which had applied to both fellows and students), meaning that non-Anglicans were now permitted to take degrees or take up fellowships. Dissenters had been calling for this repeal for half a century, and although the repeal did not entail a complete disestablishment of the Church of England, it was viewed as a significant step in the right direction.[19]

Following on from the implementation of these parliamentary reforms it was no longer obligatory to assent to the 39 Articles. Instead, an oath was enjoined on fellows of the universities of Oxford, Cambridge, and Durham (established in the early 1830s) not to teach "anything contrary to the divine authority of Holy Scripture." However, this standard became diluted over later decades of the century to the stipulation that all fellows should simply be of "good moral character."

Further debates ensued in the 1870s as to whether fellows should be able to marry. In the late 1870s, and after another government commission had reported, it was agreed that fellows should be permitted to marry after the seventh year of their tenure, although some colleges held out by imposing statutory restrictions. The church's influence was further weakened by a growing hostility to many vestigial ecclesiastical prerogatives, resulting in a downgrading of the position (and emoluments) of college chaplains.

As Owen Chadwick observed, "The liberal ideal of freedom to choose how or if to profess faith was realized in Europe between 1860 and 1890,"[20] ushering in the beginnings of that free market of opinions with which we are familiar in our modern, pluralist democracies. In time, even fellows in holy orders came to view themselves more as academics than clergymen, and by the 1890s some younger dons were even regarded as open deriders of religion. Secularism had, as it were, come of age, and by 1900 the Tractarian movement of the earlier part of the century must have seemed like a very forlorn rear-guard action.

The years 1859/60 have a particular significance in this new tendency towards secularization and provide us with a useful caesura for the periodization of the intellectual life of the Victorian era. The year 1859 saw the publication not only of Darwin's *Origin* but of John Stuart Mill's *On Liberty*, a clarion call to individual liberty against the tyranny of majorities. The year 1860 witnessed the Wilberforce/Huxley debate as well as the publication of the bland-sounding but in reality explosive *Essays and Reviews*, a collection of theological/philosophical pieces authored by six liberal churchmen and one layman.[21]

The seven essays, taking their cue from the German Higher Criticism of the Bible shunned by Oxford in the earlier decades of the nineteenth century, brought cool analytical eyes to the Bible via methods of analysis hitherto reserved for non-Christian texts. Seemingly as if to atone for Oxford's holdout against Enlightenment thinking for the best part of a century, the essays revisited the disparate textual traditions on which the Bible came to be based, rejecting older tendencies toward "indiscriminate Bibliolatry," in the phrase used by the poet Samuel Taylor Coleridge.[22]

This controversy, coming only a decade and a half after the Tractarian debates, seems to belong to another intellectual world altogether from theirs. The claustrophobic world of Oxford had given way to European voices outside its cloistered setting. Even Oxfordian Benjamin Jowett's "On the Interpretation of Scripture" advocated reading the Bible as if it were "any other book," the recommendation being a clear reference to the fundamental philosophy behind the German Higher Criticism.[23] Rowland Williams, professor of Hebrew and principal of the recently founded St. David's College, Lampeter in Wales even appended a review of the once-reviled German scholarship in his essay.

Williams, who had previously published a group of his sermons under the title *Rational Godliness*, wrote in his essay that revelation is progressive and not confined to Jews and Christians, which was likely an admiring nod to the much older work by Matthew Tindall, *Christianity as Old as Creation* (1730). Future archbishop Frederick Temple also wrote of the progressive education of the human race, a sentiment based on the late eighteenth-century German dramatist and religious thinker Gotthold Ephraim Lessing's *The Education of the Human Race* (1777) as well as his "drama of tolerance," *Nathan the Wise* (1779).

In 1860, "advanced" British thinkers no longer shrank from applying German philological scholarship to biblical texts and traditions. Temple did not renounce biblical authority, but like Lessing and Lessing's source, the theologian Hermann Reimarus, he relocated it from the text to the spirit. "The letter [of the Bible] is not its spirit," Lessing had announced.[24] Baden Powell who, tellingly, included in his essay a favorable mention of Darwin's work, counseled, following Lessing and his English-speaking successors, that the Bible should be read and interpreted only in such terms that did not contradict reason. He argued that people's faith should be strong enough to endure without meretricious miracles. Henry Wilson's essay pointed out that the teachings of Jesus were primarily moral rather than dogmatic and doctrinal. Harvey Goodwin's essay pointed to the fallibility of Mosaic cosmology, by which he meant the mythological accounts of the creation, the garden of Eden, and other *mirabilia* associated with the Pentateuch, the first five books of the Old Testament. Clearly influenced by Charles Lyell's geological work in the early 1830s, Goodwin redefined the biblical six days of creation allegorically, not as units of twenty-four hours but as geological *eras*. Genesis should not, argued Goodwin, be read as if it were a specialist geological account.

Altholz identified the common denominator linking the rather disparate contributions of the essayists as being a stress on the importance of "the still small voice" of individual conscience versus doctrinal rigidities: "The Christian revelation consists not of doctrinal propositions but of moral principles, and doctrines must conform to the dictates of conscience," he wrote. "The ultimate revelation was in conscience itself, awakened by the Church and instructed by the Bible."[25] As Altholz also noted, however, this might have seemed a rather thin religion for the multitude, and one early response in the form of a notice in *The Christian Observer* objected that in denying the supernatural elements of the Bible, the essayists were presenting their co-religionists with "a religion without a creed."[26] A related objection was that such spokesmen for Christianity were espousing rationalism without admitting as much—that is, pretending to accept Christianity and the supernatural authority of its central book and not owning up to the implications of their own arguments.

Whatever the merits and demerits of this collection of essays, its function as a marker of a changing zeitgeist is unambiguous. Nobody was more aware of the implications of the essayists' views than Oxford's bishop, Samuel Wilberforce, and it appears appropriate to give him the last word on what he took to be the repercussions of the *Essays and Reviews* affair for religionists of his day. It goes without saying that we must factor in Wilberforce's *ex officio* role as an Anglican bishop here, but we would also do well to acknowledge his superbly analytical mind. He was certainly not the kind of intellectual Luddite which his parody suggests. His extensive and well-informed review of Darwin's *Origin* and his Oxford First in Mathematics do not make him an easy target for facile mockery.

In the event, Wilberforce hit the nail on the head with regard to identifying the unifying idea linking the disparate group of essays. He wrote, "The idea of this 'verifying faculty'—this power of each man of settling what is and is not true in the Inspired Record—is the idea of the whole volume."[27] The intellectual progenitors of the volume, he continued, had been the German rationalists, themselves the heirs of the eighteenth-century English deists.[28] He finally delivered himself of a deadly Parthian shot by his observation that the *Essays and Reviews* had been welcomed most joyously of all by "brother unbelievers."

Wilberforce's barb was both pointed and prescient, putting its finger on the burgeoning naturalistic turn asserting itself in British intellectual life after mid-century. By the time Darwin published his *Origin*, the educated public had inter alia been exposed to the view that the Bible was myth, God was a projection of mankind's own needs (Ludwig Feuerbach via George Eliot's translation into English of *Das Wesen des Christentums/The Essence of Christianity*), and the universe was devoid of spiritual forces. In one sense, the *Origin* rode the tide of these new materialist currents, making its grand entrance at a particularly propitious moment, the intellectual temper in Oxford and amongst British intelligentsia more broadly having shifted markedly in Darwin's favor from that which had characterized the earlier decades of the nineteenth century.

What occurred in the circumscribed intellectual arena of Oxford was also matched elsewhere in Great Britain and Europe, as men and women adjusted their thinking to the aftermath of the European Enlightenment and the new, positivist temper of mind articulated by Auguste Comte and his followers.

In emphasizing the decisive shift in thought from the 1840s through the 1860s and beyond, the intention is not to leave the impression that there wasn't a long fuse for this explosive period of intellectual change. Indeed, in search of John the Baptists preparing the way, one could reach back to the nominalism of Ockham in the fourteenth century, to Italian Renaissance authors such as Giovanni Boccaccio and Niccolo Machiavelli, and less ambitiously, to the aforementioned deism of the Enlightenment. To quote Mark Pattison in his essay "Tendencies of Religious Thought in England 1688–1750," rationalism had long since become "a habit of thought ruling all minds."[29]

A brief look at the historical record confirms this. It had been well over a century since John Toland had published his *Christianity Not Mysterious* (1696) where he commended a *reasonable* form of religious observance *sans* miracles. In the following century, Gotthold Ephraim Lessing, influenced by the theologian Hermann Reimarus, advanced a non-literal understanding of Christianity. Europeans in the new, scientific age had begun to be more drawn to the ethical message of Christianity than to its supernatural elements. Thomas Jefferson's famous (or infamous, depending on one's perspective) project of creating a Bowdlerized version of the Gospels free of the miracle stories is of a piece with all this.

It was the intellectual endorsement of that current of thought by the seven reviewers (their voices amplified by persons outside academe), together with the happy synchronization of the publication of *Essays and Reviews* in the immediate aftermath of the *Origin*, that helped give Darwin's work the pass it needed to prevail in the face of conservative holdouts.

In modern jurisprudence there is a well-known phenomenon referred to as confirmation bias, the idea being that juries should not be unduly swayed by prior knowledge of a defendant's life and deeds. Something of this constellation of thought may have lain behind the observation of the seventeenth-century French writer Blaise Pascal to the effect that the best way of getting people to believe what you are trying to persuade them of is by getting them to think that this is what they wanted to believe all along. Such persuasion of course can occur the more readily if one's audience has been conditioned beforehand and softened up to accept the truth-value of any new group of propositions being presented to them. A prime example of this "timing is everything" principle was seen in the case of Lucretius, whose militantly materialist philosophy, after its rejection in antiquity and consignment to almost two millennia of oblivion, was reintroduced to the broader consciousness of educated Europe towards the end of the seventeenth century (though the intelligentsia of the late Middle Ages and Renaissance were aware of Lucretius's writings). Thus did *De Rerum Natura*/*On the Nature of Things* re-enter the scene on the very ground floor of the Enlightenment project, where it penetrated the consciousness of the European mind with far less hindrance than it would have faced a century or two before.

By the same token, I would argue, Darwin's repeated deferrals in the writing of his magnum opus might, from its author's perspective, have turned out to be a blessing in disguise in that the delay conspired to push back its publication by just the right amount of time to benefit from the secular turn coming to fruition in Oxford and the West generally. By 1859 many persons inside and outside the university world were actively *willing* there to be natural rather than miraculous answers to their questions about the world, an urge tailor-made for the arrival of Darwin's *Origin*.

Social Change and the Rise of Secularism

It cannot have done any harm to the success of Darwin's opus that it was packaged in a volume light on scientific jargon and laced with just the right amount of decent gentlemanly hesitancy to endear it to a British audience which might have been deterred by a showier or overly "intellectual" mode of presentation. By implicitly disclaiming airs of omniscience, Darwin avoided the vice most disliked by English readers: that of trying to appear too clever. This was likely a particular boon in his efforts to connect with persons far removed from the rarefied world of Victorian learning.

The predominantly working-class radical movement simply styled "secularism," which had its origins in earlier decades of the nineteenth century, had been gathering momentum since the early 1850s.[30] The movement has been the subject of a special study by Timothy Larsen in his *Crisis of Doubt: Honest Faith in Nineteenth-Century England*. As Larsen documents, there had been an atheistic newspaper in circulation from the early 1840s called *The Oracle of Reason* which was staffed by several "transmutationists" who accepted pre-Darwinian speculation about evolution long before Charles Darwin came on the scene and even before the publication of that work widely seen as a prelude to Darwin's *Origin*, Robert Chambers's *Vestiges of the Natural History of Creation* of 1844. There was also a weekly penny periodical, *The Plain Speaker*, which ran for a short time in 1849, plus a host of secular societies nationwide. One of the secularists chronicled by Larsen, the indefatigable John Henry Gordon, is on record as presenting papers in 1861 to secularist societies in Leeds, Bramley (the Greater Leeds area), and at a south London branch in 1862.

Learned Autodidacts

The movement was then not just a metropolitan phenomenon, and reference was routinely made to the Northern circuit of engagements arranged for speakers. In the heyday of the Industrial Revolution, Britain was considerably less London-centric than is the case today, and it was widely acknowledged that most important working-class movements and initiatives had originated and become widely diffused over what are now ruefully referred to as the old industrial heartlands (the British equivalent of the American Rust Belt).

Gordon and other speakers at such events were exceptionally well-informed across a range of disciplines. Adrian Desmond pointed out some decades ago in his *Politics of Evolution* that many new ideas in this period were typically introduced not by conservative Oxbridge dons but by medical and scientific radicals. As Larsen remarks with regard to another prolific speaker, Thomas Cooper, "If in 1850 any Britons wished to have a serious encounter with the latest modern biblical criticism, they would have been better off going to hear Cooper lecture than attending any British university."[31] This was an era in which *formal* higher education was the preserve of privilege and wealth, an exclusion which spurred many enquiring minds to alternative, autodidactic expedients.

Although Darwin provided what his later hagiographers would describe as a particularly daring set of ideas, in reality those ideas were no more audacious than ones long in circulation in the radical press and aired at the secularists' public meetings. The English upper class's virtual monopoly on higher education proved no impassable barrier to enquiring minds with access to public libraries and the pooling of collective knowledge facilitated by local learned societies. Many radicals had absorbed ideas from the *System of Nature* of the French materialist *philosophe*, Baron d'Holbach (translated into English in 1797), and Erasmus Darwin's ideas of evolution had been more or less common knowledge since end of the eighteenth century. Even in the late eighteenth century the somewhat eccentric Scottish jurist Lord Monboddo (James Burnett) had proposed a consanguinity between man and ape (although few at the time took him seriously).

The relative chronology is important here because the so-called "plebeian" writers and speakers were well ahead of the curve in terms of their readiness to accept and promote "advanced" forms of intellectual speculation. By contrast, the "upper classes" took longer to embrace these new ideas. An illuminating example of this, discussed by Larsen, is the loss of faith suffered in the 1860s (and, thus, after the *Origin* was published) by Sir Leslie Stephen, because he was no longer able to accept as literal truth the biblical account of the flood and Noah's ark.[32] Remarks Larsen, "From the perspective of plebeian radicals, what is surprising about this [loss of faith] is not the critique [of the Bible] but rather the late date. One could have gone to a freethinking hall decades earlier and heard a careful catalogue of reasons why the account of the flood, on a standard, literalist reading, could not be squared with what was known of geology, and how it was filled with a wide variety of absurdities."[33]

Confirming Disbelief

Jennifer Hecht once observed that the publication of Darwin's *Origin* was "like Christmas for ex-Christians."[34] And as we saw from the above chronology of events, there were many such among the "plebeian" skeptics of the time. For such radicals the publication of the

Origin likely elicited not so much shocked surprise as a confirmatory "Aha—I already knew the *what*; now I know the *how*!" In other words, they welcomed the *Origin* as scientific *vindication* of a religious skepticism they had come to by a different route.

Even if, like Thomas Huxley, one did not at the time think that the theory of natural selection made sense in purely scientific terms, it did provide many people with a convenient confirmation of their disbelief, in terms more scientifically detailed than any other theory of biological origins on offer. Arguably, Darwin's most noteworthy contribution to the so-called "forward march of ideas" may have been not so much to biology but to the cause of secularism, so much so that Darwin came to appear to some militant secularists as the unofficial patron saint of their cause. On one occasion in the late 1870s, two of its leading lights, Charles Bradlaugh (the first openly atheistic member of the British parliament) and his associate, Dr. Edward Aveling, a young biology professor, came to solicit support for their cause from the by-now venerable Darwin. (The bourgeois conservative turned them down.[35]) Here, quite clearly, were two men who had been primed to embrace Darwin's wholly naturalistic theory of evolution.

In Shakespeare's tragedy *Othello*, the playwright's psychological insight is nowhere more apparent than in his depiction of the negative dynamic between the eponymous hero and his villainous lieutenant, Iago. Much in the play is made to hinge on the notorious prop of Desdemona's handkerchief, which Iago had contrived to misappropriate and plant in Cassio's rooms and which Othello is duped into taking as "ocular proof" of Desdemona's adultery with the dashing young Cassio. Iago's malign stratagem works perfectly. As he had predicted, his chosen mark proved to be "as easily led by the nose as asses are." Othello, witnessing the handkerchief, tragically succumbs to his own paranoid insecurities. Iago, in a memorable description of such confirmation bias and referring to the planted handkerchief, soliloquizes in an aside to the audience, "Trifles light as air/ Are to the jealous confirmations strong/ As proofs of holy writ."[36]

In other words, Othello, already primed to paranoid jealousy, needs only the barest evidence to push him from suspicion to conviction.

The tendency is all too human, which is why in modern jurisprudence any information leading towards possible confirmation bias must be withheld from a jury to prevent them from jumping to conclusions.

And to bring us back to the main thread, in the case of Darwin and his magnum opus, by the time of its publication, there were many in England and abroad with a confirmation bias in favor of his theory, thanks to all the aforementioned who paved the way for its reception. The empirically demonstrable truth-value of the *Origin* might have been negligible, a "trifle light as air," but that mattered little to persons already primed by their prior ideological formation to accept Darwin's argument as a form of secular gospel. For the many who had already all but bidden adieu to the religion of their youth, Darwin's *Origin* came together with their prior misgivings about God and the creeds of Christendom to precipitate an even greater degree of secularist thinking, while at the same time lending these secularists a potent new tool for recruiting others to their viewpoint.

There were, however, some of the more self-critical secularists who were to experience a light-bulb moment in later life, prompting them to reassess their previous stance.

Reconversions and Dissenting Voices

Many of the secularists chosen for study by Larsen eventually returned either to the faith they had initially rejected or to some other form of spiritual orientation. Typically, they would find over time that secularism offered no positive program for people to live by. One of their number, William Hone, came to realize that materialism could not account for the totality of human experience—there must be a power *behind* matter. There was a general feeling among the reconverts that their erstwhile skepticism might have been the result of "a procrustean system of logic, an oppressively narrow definition of reason. They came to believe that human beings knew more than could be proven by such a method."[37] The reconverts, moreover, were prompted by dint of lived experience and maturer reflection to revisit "their assumption that the cause of radical politics and the working classes naturally led to an opposition to Christianity."[38]

The best-known such figure was from a modest social background, mixed with working class and lower-middle-class peers in his youth, and imbibed the lessons of secularism from a young age: Darwin's one-time comrade-in-arms and the co-founder of the theory of evolution by natural selection, Alfred Russel Wallace. Wallace must in his early life have seemed like a veritable poster boy for the secularist creed, yet he too, like William Hone, was to experience a change of heart in his maturer years.

5. DISSENTING VOICES

A LFRED RUSSEL WALLACE HAS TRADITIONALLY FIGURED AS THE man who played second fiddle to the main protagonist. Yet in many ways he has a more compelling story to tell than Darwin himself because his intellectual life came to be more varied and less narrowly focused than Darwin's.[1] In what follows I shall intercalate a short conspectus of Wallace's achievements in middle and older age (he lived well into the twentieth century and died only in 1913) because I suspect that the sheer breadth of his widening vision gave him the perspective necessary to be able to stand back from his initial work on evolution to advance a particularly thoughtful assessment of it, the sort of unsparing analysis in too little supply in Darwin's case.

Wallace's corrective to the theory of evolution jointly advanced by him and Darwin in 1858 has not, until fairly recently, received the attention it deserves, but Wallace's unsparing analysis of the theory contains an implicit challenge to us today to approach the theory not as sacrosanct scientific dogma but as an idea that should remain open to testing and critique, in our case with the added benefit of more recent scientific advances that neither Darwin nor Wallace could have foreseen.

The young Wallace's early life and career were strikingly different from Darwin's experience. He did not enjoy Darwin's material comfort and, as was the case for the overwhelming majority of Victoria's subjects, he could not rely on the safety net of family financial support in case of failure. Not for him the luxury of forgoing the world of work thanks to inherited prosperity. Instead, taking advantage of his quick wits and physical fitness, Wallace found employment as a land surveyor for a number of years, a task he performed with skill and efficiency. Although Wallace's earlier life was not without some signal incidences of ill fortune, it remained unmarked by the kind of failures and disappointments which characterized Darwin's younger years.

Born in rural Monmouthshire in 1823 (to which picturesque location his parents had been obliged to decamp from London in order to escape the high metropolitan cost of living), Wallace enjoyed only limited formal schooling but was bright and quick on the uptake. This enabled him to learn the rudiments of surveying from his brother, William, whilst the countryside around Usk and Llanbadoc gave him an ideal opportunity to teach himself botany and entomology via lovingly curated beetle collections in that unspoiled part of the county. (It is a safe twelve miles distant from the bustling industrial port town—now city—of Newport.)

As a young adult, absorbed by the wonders of nature, he decided to go off with a childhood friend of his (Walter Henry Bates, 1825–1892) to explore South America, a journey which lasted from 1848 to 1852. The voyage was financed not by his impecunious father but by the young Wallace's sale of specimens found. The journey of exploration was remarkably successful until, on the return voyage, the ship caught fire. Crew and passengers were driven from the ship by the smoke and flames. They were rescued only "after they had endured ten harrowing days and nights in a small row boat, 200 miles from the nearest land, with water seeping into the boat from numerous holes."[2] Wallace writes that he was "scorched by the sun, my hands nose and ears being completely skinned, and drenched every day by the seas and spray." Worse for his fortunes, his specimen collections were lost. "And now I began to think," he wrote, "that almost all the reward of my four years of privation and danger was lost."[3]

Thus did he return to Britain in a decidedly bedraggled state and in straitened circumstances. Nevertheless, Wallace was still able to publish accounts of his findings on the Amazon and Rio Negro, and remained undeterred by the reversal.

So it was that in 1854, undaunted, Wallace set out again on what turned out to be an eight-year voyage of discovery to the Malay Archipelago (now known less evocatively as Maritime Southeast Asia). Wallace, like Darwin, had a voracious taste for natural history, a taste that in Wallace's case had been considerably whetted by his reading of a book alerting Victoria's middle-class subjects to advances in scientific thought in the previous half century, and which was to have the profoundest influence on ordinary people's understanding of their existential situation.

Robert Chambers's *Vestiges of the Natural History of Creation* (1844),[4] thought to be more than a little scandalous by ecclesiastical representatives at the time, was originally published anonymously by a successful Scottish book publisher for fear of ecclesiastical censure as well as loss of revenue to his firm. In it, Chambers put forward an essentially materialist explanation for the creation of life on earth. He wrote (reporting on recent scientific thinking) that organic life was thought to have begun with the appearance of dry land on the face of the earth. Dry land, he explained, had emerged over eons through a combination of sedimentation and an upward thrusting of rock by forces which in time provided "a theatre for the existence of plants and animals."

Although this seemed to contradict the biblical narrative contained in Genesis, Chambers fudged a trifle by implying that the Almighty had acted by delegating creation to intermediary forces, proposing that He would not have deigned to bring forth each individual species through the exercise of "separate exertion."[5] Hence the earth and the whole solar system came not from a one-off divine creation but from natural laws which Chambers characterized as "the expressions of His will."[6]

Chambers sketched an evolutionary story from the simplest to more complex beings on the earth, all of them connected by transitional forms between species. This ladder of organic life did not appear all at once, as stated in the Bible, but developed over eons of geologic time. Decades before Darwin published *The Descent of Man*, Chambers stressed the biological continuum linking beast and man by explaining that the human fetus shows a similarity to that of an ape but that these features are "suppressed" before the baby is born and in this way the infant goes on to become a "true human creature."[7]

Wallace, reading *Vestiges* in the autumn of 1845, was much excited by its arguments. Chambers's theory convinced him of the truth of evolution, and because Chambers did not actually identify a mechanism to explain how the claimed evolutionary process worked *in practice*, Wallace was spurred on to locate such a mechanism. It was this desire which was to bring the lives and careers of Darwin and Wallace into their fateful apposition in future time. Their claim to have found the mechanism of evolution was to create a close bond between the two men. That bond was never broken (in later life Darwin pleaded successfully to procure a government pension for Wallace) despite the fact that, just five years after the publication of the *Origin*, Wallace began to have strong and publicly expressed doubts about the creative powers of the joint mechanism that he and Darwin had happened upon independently and introduced to the world jointly, natural selection working on random variation. While those doubts created a rift with Darwin at the professional level, they did not, to Darwin's credit, prevent Darwin from respecting his old ally as, to borrow the Victorian commonplace, "an honest doubter."

The apogee of Darwin's career was his publication in 1871 of the pendant volume to the *Origin* titled *The Descent of Man*, and his interests thereafter became more circumscribed than those of his peer. Whereas Darwin devoted his final years to the less controversial study of life-forms such as earthworms from the safety of Down House, Wallace spread his wings to study wider, often potentially contentious subjects in his later role as a public intellectual.

The most meritorious cause to which Wallace devoted himself in later life was his opposition to the eugenics movement. This movement, which set out with the intention of improving the human stock in ways familiar to animal farmers, started off in 1883 when Francis Galton, influenced by his cousin Charles Darwin, advocated a system of promoting the fittest in society and discouraging breeding by persons deemed less suitable. The movement achieved great popularity for a good half century before, and as is well known, it was discredited when the Nazis instrumentalized it to further their own genocidal agenda. Wallace would have none of it, finding the eugenics program detestable and ultimately ineffectual.

Wallace, it should also be noted, opposed scientific racism in its more extreme form, a form all too common in the Victorian age; and while he was not free from Victorian racial prejudice, he never had any use for the racist claims that Darwin and others employed to try to minimize the distance between apes and humans. Whether it was Wallace's working-class upbringing, or the much greater amount of time he spent among indigenous peoples than Darwin had, or an inherent strength of character which protected him from such pernicious thinking, or some combination of all three, the fact is that Wallace, unlike numerous men of science of his age even well into the twentieth century, had the discernment to reject such arguments as self-serving pseudoscience.[8]

Of course, Wallace did not always get things right. A notorious cause for which he is remembered is his espousal of the culture of spiritualism. Wallace, it should be noted, was not the only man of science in the late Victorian period to have been seduced by the cult of séances. For example, the famous author Sir Arthur Conan Doyle, a physician and student of botany, fell into the same curious category. The most that Conan Doyle's biographer could say in his defense was that Conan Doyle "remained fascinated by the uncharted potential of the human mind,"[9] and that many other grandees such as Wallace, philosopher William James, and future prime minister Arthur Balfour had also succumbed. Hence, if Conan Doyle's "interest had been restricted to social climbing, he could have done worse than join the S. P. R. [Society for Psychological Research]."[10] Hardly an entirely convincing justification for either Conan Doyle or Wallace, it may be thought, but it does nicely foreground how common this lapse of judgment was among even some of the keenest minds of the period.

Even more contentious was Wallace's opposition to compulsory vaccination, but in this case he was driven by principle, convinced that the move by the medical profession to enlist state aid to enforce universal vaccination of children under penalty of prosecution for non-compliance was a pernicious example of top-down paternalism. He also disputed statistics intended to show the unqualified benefits of such a program. In retrospect he turned out to be mistaken, the smallpox vaccination eventually succeeding in all but wiping out the disease, but it is worth noting that his motivation for resistance to both eugenics and compulsory vaccination sprang from the same antipathy to autocratic presumption of the privileged toward the less privileged.

Arguably the most impressive achievement of Wallace's older age was his contribution to the science of cosmology. In *Man's Place in the Universe* (1903)[11] and in *The World of Life* (1910),[12] he developed the idea of a fine-tuned universe, referred to today as the anthropic principle (the conception that the laws and constants of the universe are ideally fit—or perhaps fitted—for humankind, or at least for creatures like ourselves). His other cosmological work, *Is Mars Habitable?*[13] was published in 1907, when its author was already eighty-three years old. Here Wallace argued, using technical knowledge about planet cooling ratios relative to distances from the sun, that Mars was too cold to allow water to flow into the so-called *canali* (conduits) of the Martian surface and that these "canals" could not be artificial constructions made by Martian inhabitants. Life was in any case highly unlikely there, he further argued, since the atmosphere was too thin.

Wallace was not arguing in a vacuum. He was opposing an increasingly popular notion that Mars likely was inhabited. The American astronomer Percival Lowell even wrote a book entitled *Mars and its Canals* in 1905. Wallace, of course, was ultimately proved right, but the controversy remained alive for a surprisingly long time, with Lowell's fantasies about the artificially constructed "canals" of Mars not being definitively put to rest until close-up images were relayed back from the 1965 Mariner 4 flight. These images confirmed Wallace's view that the "canals" were just accidental configurations of the Martian crust.

Flawed as Wallace was in his choice of some causes, the totality of his contributions to science shows him to have been an acute and often far-seeing analyst of late Victorian and early Edwardian scientific cruxes, and it appears that the breadth of his later interests gave him an added sense of perspective with which to revisit and revise earlier phases of his scientific career.

Wallace's Apostasy

Wallace began to develop doubts about the ostensible scientific triumph he shared with Darwin a bare half decade after announcing it to him in his famous Ternate letter of 1858. At first as strong an advocate for natural selection as Darwin himself, Wallace by the mid-1860s had parted company with Darwin on the subject of the human mind with its uniquely advanced language facility. As noted above, Wallace, unlike Darwin, did not have recourse to the false belief that many indigenous peoples were little more intelligent than apes. For him, simple ape-to-human progression was no longer tenable, and he could not continue to assent to the ontological equivalence of humans and nonhuman animals supported by Darwin.

To be sure, he was at one with Darwin in concluding that brain trumps brawn. Indeed, for Wallace, the mind of man was so developed that his body was now of considerably lesser importance since the developed mind enabled humans to transcend their environment in many respects. Unlike all other species, mankind could devise clothes and fire for protection from harsh conditions, and had the resources of tools and weaponry to protect itself from foes. Hence, humans were able to survive in almost any environment, and indeed so triumphantly that the tyranny of natural selection was now effectively vanquished for humankind as regards its physiology. The issue that divided Wallace and Darwin did not concern the *fact* of human mental superiority but rather how such a priceless gift had originated.

Darwin himself had insisted that the rough-and-ready process of natural selection had no power to produce perfection. Doubtless it was for that reason that the eye, which Darwin called an organ of "extreme perfection," left him with the nagging question of just *how* it could have evolved. What then of the human mind, so far in advance of any other primate's? Wallace, as noted above, while not free of Victorian prejudices, held to a considerably less extreme form of them, and had noted in his more than a decade of fieldwork in far-flung locations of the globe that primitive tribes were not radically dissimilar intellectually from Europeans even if not (yet) their equals at the technological level. "Savages" were only required to operate in simple circumstances where their great brainpower was easily overlooked by colonial visitors who neglected to spend significant time in close association with them, as Wallace had.

At the same time, while they were undoubtedly intelligent, it was also clear to Wallace that they had no need for the capacity to learn, for instance, higher mathematics, compose complex music, or pursue various other high intellectual endeavors common in technologically advanced societies. So what was the point of their great mental powers and, more importantly, how had those powers evolved given that natural selection would not have been "called on" to enable early man to perform such cognitively challenging tasks for which there was, in their case, no need? By extension, what was the survival value of musical and mathematical abilities for primitive Europeans? These were patently not brute survival skills. How could those aptitudes have been cultivated and developed by natural selection, which favors only immediate utility, for as Darwin himself repeatedly stated, it had no power of foresight? Whence came the extraordinary powers of the human brain? For Wallace it was clear that an additional power must have played a role, something over and above what was operative in the joint mechanism of random variation and natural selection.

In March 1864, in an address to the London Anthropological Society, he first broached the subject of human exceptionalism. But he sought to explain human exceptionalism within the framework of natural selection, there arguing that as humankind's distant ancestors crept forward socially and culturally, there came a point where their social organization was sufficiently advanced that natural selection waned in its effect on humanity's physiology and instead began to act primarily on our intelligence and temperament, with natural selection rewarding variations that improved such qualities as creativity and sociability rather than strength, speed, and such.[14]

But how had humankind reached such a point in the first place? It was a decided weakness in the analysis. Five years later, in 1869, in a review of the tenth edition of Lyell's *Principles of Geology*, Wallace initiated a momentous break with the thoroughly materialistic version of evolutionary theory that he, together with Darwin, had heretofore advocated. Concluding that something more than mindless evolution by natural selection was required, he made reference to the place of an "overruling Intelligence" within the evolutionary process,[15] a statement which was notably closer to the position of Darwin's sworn enemy, the theistic evolutionist, St. George Mivart, than to materialistic Darwinism.

Specifically, Wallace argued for a special place for the human mind, concluding that some higher power must have been behind its origin. In fact, the existence of the human mind was the living proof of a divine mind, on the Platonic principle that nothing comes of nothing, since everything that comes to be requires a blueprint or original exemplar (a formulation, note, that leaves aside the First Cause, understood to be eternal and therefore uncreated). From this view he would never consent to be deterred, either by Darwin's piteous complaint that Wallace might have just "murdered… your own & my child,"[16] or by the editorializing objection of the prestigious journal *Nature* that, "to say that our brains were made by God, and our lungs by natural selection, is really to exclude the Creator from half His creation, and natural science from half of nature."[17] It was a contradiction he would not address satisfactorily until towards the end of his long life.

Wallace's worldview from the mid-1860s onwards came to rest on the acceptance that the First Cause was instantiated through teleological laws of nature. Both Darwin and Wallace were devotees of Milton's *Paradise Lost*, and I suspect that Wallace's new conception might have struck Darwin as a throwback to *Paradise Lost*, which has God presiding over orders of angels with their delegated powers, that scenario now transposed by Wallace into fitting biological terminology. Generations of readers throughout Christian Europe had read Milton as a glorious poetic consummation of the less embellished biblical accounts, although it should be pointed out that Wallace at this stage of his life was arguing his position from simple logic and observation of man and nature rather than on the authority of revelation from any of the Abrahamic faiths or indeed on the authority of any creedal religion whatsoever.

As Michael Flannery puts it, "Wallace's was a natural theology without a church, mosque, or temple."[18] He did not conceive of himself as being in the business of Christian apologetics. In fact, even though we know fewer biographical details about Wallace than we do about Darwin, we do know something about his early formation amongst lower middle-class and working-class peers, where he took a passionate interest in socialism and the education of the masses. As Rebecca Stott explains, this early formation had made of him "a freethinker, highly moral, socialist, and secular."[19]

Wallace's religion could then be described as "revealed" only in the special sense of its having been understood from the pages of nature itself. Wallace's take seems to have been that the divine power cannot be apprehended directly but can be inferred from the powers of nature. Hence the awe we feel in the presence of nature which, in the words of early twentieth-century German theologian Rudolf Otto, is an apprehension of The Holy, defined by Otto in a celebrated book as a power that instills a sense of its absolute superiority, and makes us feel our personal submission to it and submergence in it.[20]

Darwin, unsurprisingly, was of the opinion that his *confrère* had shot himself in the foot with a particularly large-caliber bullet; and his anger and disappointment over having such a close colleague let the proverbial team down was not helped when he learned that Wallace had succumbed to the contemporary craze for séances. And, indeed, Wallace's belief in out-of-body manifestations in these ritualistic gatherings may have played a role in his being willing to consider explanations for human exceptionalism that involved mind-body dualism rather than a strictly materialist framework.

One might be tempted, with Darwin, to damn Wallace's rejection of pure materialism for its association with discredited quackery, but such guilt by association doesn't follow logically. Think, for instance, of Christopher Columbus's false conviction that Earth was smaller than it actually was and Asia much farther across than it actually was. Columbus's false views emboldened him to launch out into the deep Atlantic in search of India, convinced that he would find it roughly where he instead encountered the Americas; thus, two false convictions led him to make a true discovery. In Wallace's case, his belief in the out-of-body experiences claimed of séances may have been cause for his openness to mind-body dualism, or his belief in those out-of-body experiences may have been the effect of his already having arrived at philosophical dualism by another route. But whatever the path, he came in later life to believe that mind and body were discrete if interrelated entities, leaving him the flexibility to consider the possibility that human exceptionalism involved something more than the creeping physiological advances of evolution by natural selection, and that the source of that something more was an immaterial Higher Power.

For Darwin, however, the sort of talk Wallace was now indulging in represented little more than inexcusable occultism.

It should be noted that the title of Wallace's book *Darwinism* (1889)[21] is something of a misnomer for, although the bulk of the book is a standard rendition of the Darwinian line, in the final, fifteenth chapter, titled "Darwinism Applied to Man," he reneges on the scientific concordat which he had effectively endorsed in the first fourteen chapters. Here he disagrees with Darwin's contention that mankind's higher mental faculties are to be found in inchoate form in the anthropoid apes. Man's "spiritual essence," Wallace argues, had *not* evolved by natural selection. With regard to human consciousness, all idea of "mere complication of structure" producing that result were, in the final analysis, implausible.[22]

Wallace was aware that he was lending his support to the logically troublesome position of dualism. If mankind's two natures are derived each from a different order of reality, clearly both cannot be explained in the same causal terms. It was a challenge he would not address satisfactorily until late in his long life. He resolved the conceptual confusion by dispensing with the view that natural selection had fashioned our bodies while God had fashioned mankind's *higher* faculties. He now appealed instead to the idea that an animating principle must exist behind the totality of the created order, encompassing both beast and man. In essence, he adopted a rational form of theism based strictly on the data revealed by nature itself.

Wallace's evolving conception of what could be expected of natural selection makes him something of a prophet for the modern intelligent design (ID) trend in scientific thought. The ways that his legacy has been strengthened and confirmed by important discoveries in science since at least World War II is a story of great import little adverted to in prominent accounts of the evolutionary debate, but one which is now beginning to receive the attention it deserves.

What is impressive about Wallace's testimony is the without-fear-or-favor intellectual independence it reveals in the man. He suffered no disabling sense of self-consciousness about doing a U-turn from his earlier opinions. He simply accepted the unexceptional fact that a person's opinions will change over time on maturer reflection. Wallace was, as Frank Turner once put it, primarily a disinterested student of life with no interest in orthodox posturing, even after numerous honors had been bestowed upon him later in life.[23]

The breadth and humanity of his overall vision was put in evidence in those later decades of his life when he contributed with great distinction (albeit not always infallibly) to the discussion of a number of late Victorian and early Edwardian issues of public concern. Wallace's healthy disregard for others' opinions appears to have bestowed on him the inner strength to follow the evidence where it led him. He seems not to have felt anything like the need shown by Darwin to impress public opinion or pose as a Great Man of Science, and this, I would argue, makes his testimony concerning the weakness of the theory of natural selection all the more worthy of heed.

Wallace made his change of heart look painless because (unlike Darwin) he had no interest in projecting to his peers the *bella figura* of a biological conquistador. At an intuitive level he seems to have internalized the sentiment behind the line from the eighteenth-century dramatist Oliver Goldsmith: "He who seeks for applause only from without has all his happiness in another's keeping."[24] This is suggested in a letter in which Wallace points up, with a touch of humor, how far his star has fallen in the estimate of his scientist friends committed to purely materialistic explanations in science:

> The completely materialistic mind of my youth & early manhood, has been slowly moulded into the socialistic, spiritualistic, and theistic mind I now exhibit—a mind which is—as my scientific friends think—so weak & credulous in its declining years, as to believe that fruit & flowers—domestic animals, glorious birds and insects, wool, cotton, sugar and rubber, metals and gems were all foreseen and fore ordained for the education and enjoyment of man!
> And now to <u>Tea</u>![25]

This insouciance as to others' opinion of him, and his effortless ability to bat any such opinions back at them with a wink, bestowed on him steadiness to follow the evidence where it led him without his feeling the need to dissemble or trim his position in apprehension of how others might react.

Darwin's Doubts: Lamarck Waiting in the Wings

This praise of Wallace's moral fortitude is not meant to suggest that Darwin was altogether lacking in intellectual humility or introspection. Darwin was at times afflicted by an irresolution inclining him to doubt his own theory. His frequently divided state of mind can even give him the appearance of being a devoted latter-day follower of Pyrrho, the ancient Greek philosopher who doubted whether mankind had adequate grounds for claiming *any* knowledge with absolute certainty. He certainly evidenced a Pyrrhonist streak when he felt moved to question whether his own reasoning, which in his opinion had descended from lowly and unreliable baboon ancestry, could be a dependable guide to truth at all.

Given his inability to resolve such fundamental conflicts, it was perhaps inevitable that Darwin in his later decades even began to harbor doubts about the efficacy of his pièce de résistance, his proposed evolutionary mechanism, with its claimed capacity to create the whole spectrum of the world's life-forms autonomously. Could such a positive and creative process, he asked himself, have been afforded by such a negative phenomenon as natural selection, an entity that Darwin, at the behest of well-intentioned friends, agreed might have been better described with the more modest term "natural preservation."[26]

The trouble with such a revision is that the term "preservation" is hardly well calculated to impress readers with his joint mechanism's creative capacity. Darwin seems to have sensed that if he were to go public with this preservation-not-selection concession, he would substantially dim the aura of creative power he had conjured up for his theory.

So, instead he continued to employ the term "natural selection" even while he began to cast around for supplementary theories to bolster natural selection, and to make increasing use of the Lamarckian/Erasmian idea of the relative use/disuse of organs as a co-determinant of biological development. This was exemplified when in his *Descent of Man* (1871) he found himself caught up in the challenging position of trying to explain how a line of ape-like ancestors might have transitioned into *Homo sapiens*. Among various other crucial changes, simians would, of course, have to develop the capacity for complex language. But how does one explain the rapid mental processing on which articulate speech depends? Without the simultaneous co-adaptation of the simian *brain*, how could the facility of speech, which depends on the interdependent agency of the brain *in tandem with* the specialized organs of vocal articulation, have developed by the unguided processes of natural selection?

In other words, how could chance allied to what would now have seemed to Darwin a rather emasculated form of natural selection have had the capacity to promote and synchronize operations? It is clear that in order to overcome this challenge and demonstrate the causal efficacy of his evolutionary mechanism for moving from ape to man, it would be necessary to point to a precise morphological pathway *and* a precise neurological pathway of development, ones that could somehow be selected in tandem. But the explanation Darwin advanced in *The Descent of Man* (1871) amounted to little more than some speculative guesswork, and it is particularly telling that Darwin found this problem so intractable that in order to piece together a tolerably coherent explanation, he was again forced to make recourse to Lamarckian evolution. "The mental powers in some early progenitor of man must have been more highly developed than in any existing ape, before even the most imperfect form of speech could have come into use," he wrote in the *Descent*, "but we may confidently believe that the continued use and advancement of this power would have reacted on the mind itself, by enabling and encouraging it to carry on long trains of thought."[27]

Passing quickly over the suspicious overuse of the conditional tense (one notes the number of conditional "must haves" and "could haves" in the cited words) and that rather nervous, whistling-in-the-dark phrase "we may confidently believe," it is his dependence on the unfashionable Lamarckian idea of the use/disuse of organs as an evolutionary driver which is most conspicuous here, for such a conception is clearly ancillary to where he wished to place the primary emphasis, namely on his groundbreaking idea of evolution by natural selection. Darwin may at this point have been coming perilously close to apostasy from his own theory, a crime for which Alfred Russel Wallace had been arraigned in the mid-1860s. At the very least, we sense that Darwin's trumpet was giving forth a less certain sound in 1871 than it had in the 1850s.

Some later evolutionists also found it difficult to lay the ghost of Lamarck entirely to rest, and this despite the fact that twentieth-century advances in knowledge of Mendelian genetics appeared to rule out a Lamarckian evolutionary pathway.[28] The lure of Lamarck was exemplified most strikingly in the case of early twentieth century Viennese biologist Paul Kammerer and the unhappy affair of the "midwife toad." In the early decades of the twentieth century, the highly regarded Kammerer made the following astounding claim, which I cite here in the words of his modern biographer:

> Kammerer took a type of toad that is one of the few amphibian species that mates on land and forced them to breed in water. As a result, the males developed nuptial pads, which are regularly

found on other male toad species. These nuptial pads help the male toad grasp the slippery female when they copulate in water. Kammerer asserted that not only was he successful in inducing the development of nuptial pads but also that they were passed on to the next generation.[29]

Unfortunately it was later revealed that the mating pads had been faked by the use of dark ink stains, and a short time after this discovery the unfortunate Kammerer was driven to take his own life.[30]

There is no doubting the attractiveness of the Lamarckian theory of acquired characteristics and their supposed heritability. It promotes the comforting notion that parents can pass down not only their wealth and property to their progeny but also the benign results of their own physical efforts at self-improvement.[31] It is without doubt a more inspiring philosophy than is Darwinism; but such sentimental considerations were an extraneous issue to Kammerer, who simply found himself unable to accept the postulation that natural selection possessed the efficacy claimed for it by its originator.[32] Hence, not unlike like Darwin himself, Kammerer was driven to seek a Lamarckian supplement.

A similar scenario arose in the case of the late polymath Arthur Koestler, who was also moved to flirt with Lamarckian ideas out of a dissatisfaction with Darwinism as a defensible evolutionary pathway.[33] Koestler felt that Darwinian mechanisms could be at best only *part* of the picture, claiming rather like the older Darwin that "there must be other principles and forces at work on the vast canvas of evolutionary phenomena."[34] He cited the veteran Ludwig von Bertalanffy on this point, Bertalanffy having been one of the distinguished contributors to the interdisciplinary conference of internationally renowned scientists and scholars organized by Koestler in Alpbach in Austria in 1968. The proceedings of that symposium were compiled in a volume entitled *Beyond Reductionism*, and there we see Bertalanffy remarking, "If differential reproduction and selective advantage are the only directive factors of evolution, it is hard to see why evolution has ever progressed beyond the rabbit, the herring, or even the bacterium which are unsurpassed in their reproductive capacities."[35]

In light of this much-reported deficiency in the explanatory power of Darwinian theory, Koestler was drawn to Lamarckism as at least one possible means of plugging the Darwinian gap.

It should be stressed that Darwin's hesitancies and diffidences were real and not the product of false modesty, as evidenced by the five emended editions of the *Origin* which followed in brisk succession in the thirteen years following the first edition of 1859. In the later editions he did his honest best to integrate criticisms made by others, to which he always remained acutely sensitive. Darwin had always conceded that he was advancing his present theory until such time as a better one might present itself. It seems that what at first blush might be mistaken for mere gentlemanly humility was in fact meant in good earnest.

It is therefore more than possible that Darwin would have taken a keen interest in many modern findings as a means of complementing and enriching his own work, and the last few decades have in fact provided an intriguing addendum to the whole Darwin/Lamarck saga. The idea of heritability "beyond genes" is now regularly studied under the umbrella rubric of epigenetics;[36] and although results of this recent research have proved resistant to definitive interpretation, modern scientific advances have at the very least confirmed the worries of Darwin and the suspicions of Kammerer and Koestler that Darwinian explanations centered on random variation and natural selection could not possibly represent the *whole* story.[37]

Squaring the Circle: Darwin's Peers

Darwin himself was not totally convinced by his own theory. Neither, as we have seen, were many of his peers, including such friends and scientific luminaries as the aforementioned Charles Lyell. The overall impression one gains on recording the reaction of his inner circle is that most of them were willing to assent to Darwinism only in part. Modern Darwin devotee Richard Dawkins conceded as much in an interview with the BBC's Mark Urban on *Newsnight* in late 2019. Dawkins ruefully told Urban that many of Darwin's peers assented to Darwinism when speaking with one side of their mouth while telling a different story with the other. *Plus ça change!* Some late Victorians were even driven to exercise what Gillian Beer politely called a "creative misprision," which she defined in these terms: "One of the persistent impulses in interpreting evolutionary theory has been to domesticate it, to colonise it with human meaning, to bring man back to the centre of its intent."[38]

In the final decades of the nineteenth century and well into the twentieth century, Darwinism remained essentially a broad church in which a variety of opinions were debated. Nor was this plural understanding of Darwinism peculiar to Britain. A similar array of minority verdicts is also discernible in the reception of Darwinism in Europe. In Germany, Ernst Haeckel's ideas were to evolve to a surprising degree from his youth to his older age. Though he had once rejected the idea enshrined in the term *Naturphilosophie* (which invoked a mystical belief in a World Spirit), he rather curiously denied later in life that his monism was coterminous with materialism. That gnomic contention might have remained puzzling had not Haeckel expressed his later conviction that both matter and sentient life were *beseelt* ("animated" or "endowed with soul").[39] As Alfred Kelly put it, "In spite of himself, Haeckel came close to succumbing to teleology.... His view of nature resembled a giant work of art, almost yearning for the creator he kept begrudging it."[40] Haeckel's Darwinism thus appears to have modulated into a form of pantheism, a belief in that very "All-Seele" (spirit of the universe) he had once so firmly rejected.

His contemporary, Wilhelm Bölsche, who did much to popularize Haeckel's views, even came over time to view science as a form of second Reformation, whose universal laws represented a form of progressive revelation to humanity. For Bölsche, the Bible—read in its proper allegorical or what a later generation might call its "demythologized" sense—was fully compatible with science. He viewed the Bible as he did other myths and folklore, as a vital yet primitive expression of the human spirit, later systematized by science on a higher level of expression.[41] For him science was the *consummation* of the Old and New Testaments; hence the most devout position for modern man was to espouse the scientific *Weltanschauung.* Bölsche could even write that Darwin had been working in the temple of Nature, picking up where Jesus Christ had left off!

What began with many as unresolved tensions was over time resolved in favor of cognitive clarity as individuals began to analyze Darwin's data and process it in ways which satisfied their own internal sense of logic. For scientists a century ago or more, Darwinism was not the nihilistic monolith it has more recently become, not least because for scientists of that era, commonsense rationality would often peek through unbidden. Many concluded that the postulate of natural selection would not fly as a stand-alone hypothesis.

The question therefore arises of why Darwin's view persisted while the defection of Wallace and the resistance of Darwin's scientific peers have been disregarded or airbrushed out of history. Predictably, the quintessentially English subject of class has been invoked to answer this question. Sociologists of science often point to the fact that the progress of scientific ideas advances in part as a form of social process, and Darwin, unlike the impecunious and socially less well-placed Wallace, was fortunate to have an upper-middle-class support group to promulgate his ideas. How convincing is this thesis as an explanation for Darwin's greater success? I have argued that the major role in the acceptance of Darwinism depended not so much on social factors but on the truly seismic changes in attitudes to religion experienced by all classes of society by the middle of the nineteenth century. However, this does not mean that social factors played no part at all. How might those factors be characterized and, if possible, quantified?

Darwin's In-Group

There are indications that over time Darwin gained something of the status of a cult leader (in a benign sense). After all, there cannot be many natural scientists who have inspired a follower to write a fulsome, 50-page poem in their memory, but after Darwin's death in 1882 this is precisely what a younger acolyte, the naturalist George Romanes, did, commemorating him in a poem titled with lapidary simplicity, "Charles Darwin: A Memorial Poem."[42] There is ample evidence in Darwin's voluminous correspondence with both indigenous and overseas scholars, as well as in the pilgrimages to Down House he inspired from his old-boy network of former college friends and tutors, that he had an enviable gift for friendship, even to the point of inspiring forms of fraternal love.

Only on the assumption of such personal magnetism can we understand such things as his limitlessly supportive inner circle meeting regularly to discuss matters of personal and professional interest with him. The severe-looking photographs of the bearded patriarch that have come down to us clearly give few hints of the warmth and charisma he must have projected to inspire such admiration and affection. Romanes's poem, which set off the high honor already accorded to Darwin in his burial in the north aisle of the nave of Westminster Abbey near to Sir Isaac Newton, might have suggested to some an aura close to sanctity, or at the very least a symbolic assumption into something like a scientific empyrean.

Darwin's circle was not just a small world; it was an enclosed, exclusive world. To those acquainted with modern Britain, a country which frowns on nepotism and cronyism (at least officially) and which has opened itself up to meritocratic selection procedures and the importation of foreign talent, it is rather surprising that the same close-knit cast of characters keeps popping up again and again in the drama of Darwin's life, almost as if in a soap opera.[43] Clergyman and botanist John Stevens Henslow (1796–1861)[44] would regularly hold soirées at his home which would be attended by Darwin and Darwin's Cambridge mentors, William Whewell and Adam Sedgwick, the latter having been Darwin's companion on a number of geological field trips when Darwin was younger and in better health. Henslow's daughter was later to marry one of Darwin's closest friends, the botanist Joseph Dalton Hooker. It was Henslow who recommended Darwin for the *Beagle* expedition in the early 1830s and again Henslow who chaired the famous Oxford debate in 1860 where Bishop Samuel Wilberforce squared off against Darwin's "bulldog," Thomas Henry Huxley.

Despite his reservations about Darwin's ideas, Henslow's avuncular relationship with Darwin bade him always do his best to protect Darwin from harsh criticism.[45] The same was the case with Adam

Sedgwick. He disagreed with Darwin's ideas in the *Origin* so radically that, far more in sorrow than in anger, he once described Darwin's ideas in a confidential letter to paleontologist Richard Owen as being at one and the same time saddening and risible. For him, his erstwhile protégé was a teacher of error instead of being an apostle of truth.[46] Notwithstanding these reservations, he remained on commendably friendly terms with Darwin for the remainder of his life.

The recipient of this amount of indulgence from his friends clearly had every reason to feel secure in the knowledge that he commanded a supportive in-group whose loyalty he could depend on. So it was that in 1856, at a hush-hush meeting he convened at Down House, he took soundings with Hooker and Huxley as to how best to proceed with his heretofore secret ideas concerning evolution. Huxley had condemned ideas similar to those of Darwin when they had been presented in Robert Chambers's *Vestiges of Creation* (1844), and he harbored grave doubts about Darwin's special theory of natural selection. However, he held to the more general idea of universal common descent over against belief in special creation, and for him that conviction, shared with Darwin, was apparently common ground enough for him. He immediately volunteered to defend Darwin's ideas, being more than willing to take Darwin's corner against the high authority of leading paleontologist Richard Owen.

In the words of Iain McCalman, alluding to the fact that so many of Darwin's intimates were part of an old sea-dog confraternity who had made voyages of scientific discovery of their own, Huxley had come aboard and "joined Darwin's fleet."[47] Huxley might have been, in Peter Bowler's phrase, a "pseudo-Darwinian" (that is, a believer in evolution but not natural selection) yet he would not hear a word said publicly against Darwinism in any of its facets.

Darwin in America

There is no getting away from the socially parochial aspect of English life in Darwin's day. The same names recur in the Darwin story simply because debate about matters of high import at the time were debated and largely decided by an "upper crust" of ex-"public"[48] schoolboys and

Oxbridge graduates. These persons would typically not even meet, let alone converse with, members of "lower" social classes (except in trading transactions) because it was tacitly accepted that it was only the views of the social elite which *counted*.

It is a testimony to the honor among the Oxbridge members involved that the working-class Alfred Russel Wallace got the consideration he did when he inadvertently threatened to pip Darwin at the proverbial line with his letter from Ternate outlining his discovery of a theory of natural selection and asking Darwin's advice on how to proceed. From our perspective it's easy to regard the affair with a jaundiced eye—Darwin failing to acknowledge that his decision to delay was his own fault; that his timidity had sacrificed his right to priority in advancing the singular theory; and that he had no right to crowd in with Wallace for the announcement at the Linnean Society, the younger man not only having made the discovery independently but also having daringly set his mind on publishing the revolutionary theory in all haste. And while there is an argument for such a critical assessment of Darwin and his clique's response to the Ternate letter, it is well worth putting ourselves in their shoes. Darwin and his close-knit circle of socially well-placed friends could have simply frozen Wallace out and moved forward with an announcement of Darwin's theory with little more than a footnote for Wallace; and likely no one among the intellectual elite would have batted an eye.

What there was no question of was Darwin's being left out in the cold. He and his friends were simply too well-placed in British society for that to happen.

A less class-bound kind of protectiveness was shown to Darwin in the United States by Professor Asa Gray of Harvard University, perpetually vigilant in defending Darwin from the criticisms of someone with whom Darwin had already crossed swords some two decades previously over the Glen Roy dispute, the Swiss-American biologist Louis Agassiz.

Agassiz wielded considerable influence on the North American continent. Americans had welcomed the brilliant Swiss to their shores upon his appointment to a chair at Harvard and looked to him for guidance in matters scientific. Agassiz had warned his American audiences and readers against the acceptance of what was known, after the work of Lamarck and Robert Chambers, as the "developmental theory." For this, thought Agassiz, constituted a mischievous and unscientific conjecture. Agassiz still held firmly to the decidedly pre-Darwinian position that species and the higher taxanomic units were indicative of a divine intelligence.[49] His essentially Platonic view of the world (Agassiz had been taught by the French naturalist Georges Cuvier) accorded the invisible order a unique place in accounting for the purposiveness of the sentient world. His was a resolutely vitalist and Christian cosmogony.

As for Gray, he did his best to make sure that Darwin had a fair hearing and adequate defense against Agassiz's barbs.[50] Indeed, seeing that Agassiz was a major hurdle to the acceptance of Darwin's ideas in America, Gray went on something of a charm offensive, going out of his way to defend Darwin against charges of atheism and "rank materialism" and carefully framing his tactical defense in the kind of language he knew Agassiz himself was accustomed to use. In Agassiz's own copy of Darwin's *Origin*, he had written in a margin, "What is the great difference between supposing that God makes variable species or that he makes laws by which species vary?"[51] Gray batted this formulation back at Agassiz as a means of defending Darwin, arguing that a conception of God as the first cause behind adaptation and variation was no less theistic than was supposing that a supreme power had created an immutable world in an original spurt of creation.[52]

According to Gray's framing of the matter—which sat a bit loose on Darwin's precise formulation—God could be postulated as an immanent force in Nature. Such an interpretation shrank the distance considerably between the divine and mankind, since on Gray's understanding, God was omnipresent in the evolutionary process, a tutelary deity or, in modern parlance, a cosmic micromanager superintending the adaptive contrivances of Nature. Such a view of evolution was an uneasy fit with the kind of serendipitous process emphasized by Darwin—i.e., *natural* selection working on *chance* (not purposeful) variations. Gray, however, was writing as an advocate, and to that end he defended Darwinism in a way that he well may have known Darwin could not personally embrace (even if, as it turns out, Darwin was more than happy to bankroll Gray's pamphlet to aid the cause of evolutionary theory). For Gray, it may have been a case of wishful thinking, of remaking Darwinism into a theory more to his liking. Or perhaps it was a case of attempting to mollify Agassiz and his followers by playing up the theistic hints in the *Origin* and aggressively downplaying the overridingly naturalistic tenor of the work.

Constructing Darwin's Propaganda Offensive

Such proselytizing by Darwin's allies played a significant role in gaining a favorable hearing for the *Origin*. His friends were well placed to exert influence in the educational and academic spheres. The journal *Nature* was founded in 1869 in part as a vehicle for spreading the Darwinian message, and the X-Club (founded by the indefatigable Huxley) became the venue for Darwinian stalwarts to meet and discuss the progress of their "campaign." For his part Darwin exploited his status as a Fellow of the Royal, Linnaean, and Geological Societies to cultivate friendships with influential members of the scientific community, always giving priority to the role of his co-adjutants in his enterprise.

These factors support Peter Bowler's thesis that "it was through persuasion and through success in the politics of science that Darwinism came to dominate British biology,"[53] a contention seconded by a recent work focused on the subject, *Darwin's Apostles*.[54] In 1866, for instance, Joseph Hooker delivered lectures favorable to the Darwinian viewpoint at the British Society for the Advancement of Science. John William Draper (who had spoken in favor of Darwin at the Oxford Debate in 1860) was another apostle who enthusiastically championed the Darwinian argument in the US. The most fervent of the apostles was of course the proverbial bulldog, Huxley, who kept Darwin's name to the forefront for fully two decades by a series of talks and contributions to journals.[55] He also supported the candidature of like-minded persons to scientific posts and was in this way able to impose a de facto ideological bar on those who did not share views conformable to those of members of the X-Club.

Friends Pulling Punches

Like Wallace, who over time began to see through and beyond a conjecture which for him had lost any appearance of logical coherence, others sympathetic to Darwin and Darwinism came to harbor misgivings about the theory, but their reverence and love for Darwin bade them exercise a measure of tact in deference to Darwin's feelings. Nevertheless, although prepared to temporize with Darwin in the name of gentlemanly solidarity and friendship, many contrived tacitly to move closer to Wallace's version of theistic evolution. They appeared to be at one with Wallace in renouncing the "purer" Darwinian faith, yet contrived by tactful intellectual maneuvers to signal their solidarity with Darwin and so uphold the group's esprit de corps. They were, in effect, humoring the great man. Such trimming on the part of so many of his peers is one indication that hardline Darwinism was far from universally accepted by later Victorians, even among the intellectual vanguard. What *was* accepted was as often as not an artful fudge of the Darwinian position, and this fudge may explain why Darwinism was not more directly opposed. Having assumed protean form, it was no easy target to pin down.

To isolate and quantify the factors which ultimately led to the acceptance of Darwinian ideas is not a straightforward task. The theory has been much modified, festooned with revisionary patches akin to the epicycles employed to prop up geocentrism—patches that continue to multiply to the present day as old patches fail and new patches are ordered up. Therefore, even if one remains convinced that somewhere at the core of Darwin's theory shines the truth immutable, one must grant that something more has carried it past the rocky shoals of uncooperative evidence into contemporary textbook orthodoxy.

My candidate for that *something more*, at least as regards the early decades of the theory, should by now be clear: Darwin's *Origin* benefited from what in modern terms might be described as a slickly coordinated publicity campaign orchestrated by socially well-placed supporters. One must also include the scarcely quantifiable element of Darwin's personal charisma. Such a form of personal appeal clearly transcends the issue of class; yet it succeeded in inspiring a form of discipleship from many of his supporters. That factor, too, must be thrown into the mix in explaining Darwin's influence, and taken together with that more general shift in religious attitudes which had occurred in Britain between the 1830s and 1860. By the time of the famous Huxley/Wilberforce debate in Oxford in 1860, the times they were definitely a-changing, and to such a degree that I would hazard the guess (as indicated above) that, had Darwin published his *Origin* in the 1840s (as some friends had urged) rather than in 1859, it might not have achieved the same positive resonance it did. It was the secular turn in the national mood *plus* the author's charisma *plus* the fact that a socially well-placed Darwin was able to call on the support of well-placed friends and foreign colleagues—all of these factors coming together in a powerful conjunction to ensure the success of *The Origin of Species*.

However, as we have seen, this did not mean that there were no voices of dissent. Not all agreed to every jot and tittle of Darwin's postulates. Even Thomas Huxley had severe doubts about the validity of natural selection, while Charles Lyell too had critical views regarding the methodology Darwin used and the conclusions he drew from his working methods. And of course there entered that great elephant in the room, the defection of the theory's co-formulator, Alfred Russel Wallace himself. These and other voices stood in the way of Darwinian materialism utterly sweeping the field.

There was also this. Even as the secularist drive sketched out in Chapter 3 was gaining momentum, a countercurrent was gaining strength, a re-envisioned spirituality in Britain. To this large topic we turn in the following chapter.

6. NATURE'S GOD

To use the term "demythologization" today is to trigger associations with the mid-twentieth-century German theologian Rudolf Bultmann and his thesis that many mythological accretions of the Bible were in their literal form no longer credible or even comprehensible to people of the modern, scientific age. Such miraculous biblical accounts as those concerning Noah's flood, hell, the virgin birth, and Jesus walking on water were, Bultmann urged, not to be believed literally but interrogated as to their metaphorical or allegorical meanings: they were to be *demythologized* for the modern, scientific age.[1]

Yet although it fell to Bultmann to give lexical currency to the commonly referenced term (*Entmythologisierung*), he was very far from being the first to pursue the project of so-called demythologization. Even medieval theology embraced allegorical and metaphorical readings of the Bible under the rubric of what was termed the fourfold exegesis of scripture, and some early Church Fathers, including the great St. Augustine of Hippo, did not view the days in Genesis 1 literally as units of twenty-four hours.[2] The Hellenistic Jewish scholar Philo of Alexandria, whose life overlapped with that of Jesus Christ, was an early practitioner of the allegorical method of biblical exegesis. Philo and other Alexandrian allegorists believed that seemingly unremarkable everyday events or narratives might nevertheless contain a deeper, symbolic meaning expressive of an eternal, Platonic idea. Such narrative events contained within themselves a hidden dimension of meaning which entirely transcended their historical and chronological situation.[3]

Like the Greek philosopher Plato, Philo thought the allegorical or spiritual meaning of events to be more significant and "real" than the quotidian narratives which provided the point of departure for his exegesis. For him and his Alexandrian peers, the search for hidden meanings was to become an ingrained habit of mind. In the nineteenth century an informal impetus towards the revival of such modes of interpretation was provided by the scientist whose work Darwin was to take as the methodological blueprint for his *Origin of Species*, namely Sir Charles Lyell. Both as a geologist and as a churchman, Lyell was to find himself in the vanguard of the movement tending eventually towards twentieth-century demythologization.

Lyell

As observed previously, against the schema propounded by the catastrophists, Lyell advanced the more prosaic and decidedly non-supernatural theory of geological uniformitarianism, the non-divinely directed development of geological features over large tracts of deep time, development driven by the inscrutable effects of terrestrial and subterranean pressures. In his wholly naturalistic interpretation of the evidence, Lyell differed somewhat from his geological predecessor James Hutton. Hutton's version of uniformitarianism, first advanced in his *Theory of the Earth* (1788), accommodated an *ultimately* divine origin for nature's designs. Conceiving of the terrestrial environment holistically as an interconnected whole in a way reminiscent of the modern-day Gaia hypothesis,[4] Hutton had argued that the purpose of the earth was to maintain vegetable and animal life for the benefit of mankind. Animal life was dependent on plant life, plant life upon soil; soil originates through disintegration of solid rock, and so on.[5] In Hutton's view the earth as a whole was ultimately a divine design.

On account of Darwin's religious doubts, he was inclined to side not with Hutton but with Lyell's strictly naturalistic interpretation of the geological record. As we saw, however, serious disagreement between Darwin and Lyell arose from Lyell's conviction that Darwin was misapplying the analogy of the inanimate sphere of geology to the infinitely more complex sphere of organic life. Additionally, the two men held to very different conceptions of how religious faith should be defined and lived out.

Darwin found his religious faith undermined by what he saw as the undirected and godless processes of both geological and biological evolution. Inclining to a literal reading of the Bible, he concluded that the Old Testament with its creation week, the Tower of Babel, the introduction of the rainbow as a sign of hope following Noah's flood, and other such marvels were simply presenting a false picture of the world's history.[6] Miracles of any sort he regarded as incredible in the light of what science had discovered about the unvarying laws of nature. He was especially offended by the idea of eternal hellfire since he concluded that this would condemn his free-thinking father, brother, and many of his friends to undeserved torment in the afterlife,[7] and so eventually his doubts grew to encompass the truth status of Christian revelation in the widest sense.

Nick Spencer has contended that Darwin's faith had from the start been more rooted in the dogmas and externals of the Christian faith than in an inward experience of the divine.[8] In modern parlance his faith might be termed a form of Christian literalism or fundamentalism over against an experience of personal transformation. Lyell's faith, on the other hand, was of a quite different order. Holding to the view that science of whatever stripe was unable to penetrate the deeper mysteries of life, he had, in the words of Robert Reiss, "no difficulty thinking that the universe was far older than conventional religion asserted."[9] For Lyell this did nothing to undermine the *spirit* of his Christian faith, in the sense that the term spirit had been developed by the German theologian Hermann Reimarus, the dramatist Gotthold Ephraim Lessing, and others in later eighteenth- and early nineteenth-century theological thought.[10]

Lyell felt that Darwinism did not and, in the very nature of things, *could not* provide convincing answers to some of the problems it claimed to solve, because the existential questions Darwin attempted to confront and adjudicate through his theory lay beyond the proper domain of empirical science and its strictly delimited methodological parameters. Lyell might perhaps be viewed as an early exponent of what Harvard biologist Stephen Jay Gould would later formalize in his conception of the "non-overlapping *magisteria*" of religion and science (i.e., distinct specialisms with their own epistemological boundaries). Further, Lyell, like many British naturalists even by the early nineteenth century, disbelieved the story of a universal flood because he had begun to interpret Genesis allegorically. Indeed, many of Darwin's friends and colleagues, such as Adam Sedgwick, John Stevens Henslow, and even the pious skipper of the Beagle, Robert FitzRoy, were open to metaphorical readings of the Pentateuch (the first five books of the Old Testament). Darwin himself, on the other hand, remained stubbornly behind the curve of such developments in Christian thought.

Darwin and Lyell then may be viewed as standing on either side of a fault line that divided and to some extent continues to divide the Christian world. Darwin rejected allegorical readings and remained committed to the view that the authors/redactors of the early chapters of Genesis meant for that material to be taken literally. On the other hand we have Lyell representing an important straw in the wind blowing in the direction of future developments in theological thinking, and eventually to Bultmann and the many theologians who took their cue from Bultmann in the later decades of the twentieth century.[11]

Between Fundamentalism and Modernism

The scriptures undoubtedly represented the master narrative and common cultural currency of the Victorians. Literary critic Northrop Frye described the Bible as "the great code" for understanding works of Victorian creative literature.[12] So tenacious was the Bible's hold that even Charles Bradlaugh, the first atheist member of the British Parliament, remained haunted by the psalmist's reproach, "The fool hath said in his heart, There is no God."[13] Bradlaugh always thereafter chose his words carefully, claiming that he did not assert that there was no God since to deny what is unknown would be absurd. He would also, as incongruous as it might now appear, often resort to what was known as "proof-texting," this being the use of biblical passages to support and clinch arguments; and he even went as far as to express admiration for the Bible as a "vast residuum of moral beauty and grandeur."[14]

Yet the German Higher Criticism, which cast doubt on the miracles and other content deemed mythological in the Bible, was to gain considerable traction. It was for this reason that in 1902 the agnostic William James gave special prominence in his *Varieties of Religious Experience* not to the Christian Bible but to his extended expositions of the key words of his title, namely Varieties (in the plural) and Experience (as opposed to doctrine). James thought that religious experiences, not creeds, were the more important factor in people's spiritual lives. All else he deemed to be secondary, supererogatory outgrowths—theologies, philosophies, ecclesiastical organizations, and the like.[15] As a result of the increasing trend towards demythologization, a great many modern theologians now regard the Bible as an important spiritual guide but one to be consulted critically with regard to what they regard as its mythological features and errors. Even some who would consider themselves as leaning theologically conservative see evidence in key passages of a more poetic/symbolic mode of discourse than had previously been supposed.

But Darwin, it appears, was innocent of all such theological and hermeneutical sophistication, a fact that may have made him more vulnerable to festering religious doubts, misgivings that would torment him throughout his life. This was in fact pointed out to him at the time by his wife, Emma, when she counseled him in a letter not to expect the same standards of literal truth in religion as he sought in science.[16] But her words, clearly attuned to up-to-date theological thinking as they were, failed to strike a chord with her husband.

It is fortunate that the Gordian knot that was the nineteenth-century literalism-versus-liberalism dilemma was for many to be at least loosened by a coeval of Charles Lyell, the poet William Wordsworth (1770–1850). At the same time that Lyell was conducting his geological researches for publication, Wordsworth was producing an inspirational body of poetic works in which he developed a distinctive philosophy of nature that could be characterized as demythologized natural theology. His unique vision depended not on conventionally revealed religion but on the revelations of nature itself. He did not commit his many devotees to a catechism or oblige them to assent to any form of creed. In a short space of time the Wordsworthian influence was to form a complement to more conventional Christianity in the hearts and minds of many Britons, and in such a way as to buttress conventional religious observance rather than threaten it. In particular, it provided a philosophy of nature inherently opposed to Darwinian notions and played a part in encouraging public resistance to both *The Origin of Species* and to its more radical metaphysical implications.

The Third Way: Wordsworth and the Theology of Nature

Alvar Ellegård's minutely documented study of how the British public reacted to Darwin's *Origin* has rightly become a classic in the field.[17] In that volume he details the various ways in which newspaper readerships were inclined to reject Darwinism, but he leaves out of account the influence of Wordsworth, that other figure who had made a strong claim to a knowledge of nature's secrets. The view of nature as a source of spiritual revelation held by the poet laureate of pre-Victorian and early Victorian England stood in implicit but very conspicuous opposition to the opinions of Charles Darwin. It was Wordsworth who, despising the "gaudy and inane phraseology" of Erasmus Darwin and his poetic generation, announced in his Preface to the *Lyrical Ballads* (1798) that he would address the subject of mankind's relationship to external nature in the unvarnished mother tongue spoken by all British subjects. The poet's clear voice has sounded down the generations and still continues to shape Britons' perceptions and experiences of their natural surroundings. He stands out for many as a prophet of nature rather than as a mere poet.

The abstract concept of natural theology was not unfamiliar to Darwin himself following his study of William Paley's *Natural Theology* in his university days. There is even some evidence that the younger Darwin's explorations brought with them a more direct, quasi-Wordsworthian awareness of Nature's divine disclosures than he could have gotten from reading Paley alone. This was not to last, as Darwin explains in his autobiography:

> Formerly I was led by feelings such as those just referred to, (although I do not think that the religious sentiment was ever strongly developed in me), to the firm conviction of the existence of God, and of the immortality of the soul. In my Journal I wrote that whilst standing in the midst of the grandeur of a Brazilian forest, "it is not possible to give an adequate idea of the higher feelings of wonder, admiration, and devotion which fill and elevate the mind." I well remember my conviction that there is more in man than the mere breath of his body. But now the grandest scenes would not cause any such convictions and feelings to rise in my mind. It may be truly said that I am like a man who has become colour-blind.[18]

Elsewhere in the same work he further details his changed state of mind, where there is even a hint that his earlier outlook may have been at least cultivated by his reading of Wordsworth, among other poets:

> Up to the age of thirty, or beyond it, poetry of many kinds, such as the works of Milton, Gray, Byron, Wordsworth, Coleridge, and Shelley, gave me great pleasure, and even as a schoolboy I took intense delight in Shakespeare, especially in the historical plays. I have also said that formerly pictures gave me considerable, and music very great delight. But now for many years I cannot endure to read a line of poetry. I have tried lately to read Shakespeare, and found it so intolerably dull that it nauseated me. I have also almost lost any taste for picture or music.... My mind seems to have become a kind of machine for grinding general laws out of large collections of facts.[19]

One such general law he surely had in mind was his theory of evolution, fueled by that brutal struggle for existence unredeemed by any higher purpose. How far removed this is from Wordsworth, whose conception was that of nature as a source of spiritual guidance.

Wordsworth's biographical and spiritual trajectory in life was notably different from Darwin's. After a brief flirtation in France with

Revolutionary ideas,[20] his permanent abode became the English Lake District, a then-remote location which gave him ample opportunity to explore the natural world. Poems such as the "Prelude," "Excursion," "Ode on Intimations of Immortality" and "Tintern Abbey" might almost be read as object lessons to accompany William Paley's formal expositions of natural theology. For with Wordsworth, as J. R. Watson once observed, we encounter "structures in the poetry which are akin to fundamental and primitive patterns of belief," pointing to the relevance of Mircea Eliade's *The Sacred and the Profane* as a hermeneutic tool to unlock some of Wordsworth's meanings.[21]

For ease of reference I give here a short sample of just one of the passages in "Tintern Abbey," where Wordsworth relates a kind of epiphany he experienced as he beheld again the leafy ambience of Tintern in the Wye Valley, and specifically his memories of the abbey's sylvan environs on the occasion of a return visit there after an absence of five years:

Nor less, I trust,
To them [his memories] I may have owed another gift,
Of aspect more sublime; that blessed mood,
In which the burthen of the mystery,
In which the heavy and the weary weight
Of all this unintelligible world,
Is lightened:—that serene and blessed mood,
In which the affections gently lead us on,—
Until, the breath of this corporeal frame
And even the motion of our human blood
Almost suspended, we are laid asleep
In body, and become a living soul:
While with an eye made quiet by the power
Of harmony, and the deep power of joy,
We see into the life of things.[22]

Wordsworth gives voice to the aesthetic and emotional dimension of human experience, offering added validation to the philosophical arguments expounded by Paley in his *Natural Theology*.

A Wordsworth Disciple: William Hale White

Many reported being converted to a Wordsworthian view of the world after reading him. Most striking of many similar reactions was the unequivocally spiritual response reported by William Hale White, whose (thinly concealed) autobiographical novel of 1881 focused on its author's triumph over his former religious doubts.[23] White said that the experience of Wordsworth's *Lyrical Ballads* gave him the strength to go on living in a world in which the old verities had come under progressively heavier attack from the sciences of geology, astronomy, and evolutionary biology. As the narrator of the work explains, God was for him suddenly no longer remote and notional, but had finally descended from the invisible realm to take up a new abode in Nature itself. He now appeared as an "actual fact" before his very eyes. God was now at last a presence he could, to borrow a Wordsworthian phrase, feel upon his pulses. The moment of conversion experienced deserves to be cited *in extenso*:

> During the first two years at college my life was entirely external. My heart was altogether untouched by anything I heard, read, or did, although I myself supposed that I took an interest in them. But one day in my third year, a day I remember as well as Paul must have remembered afterwards the day on which he went to Damascus, I happened to find amongst a parcel of books a volume of poems in paper boards. It was called "Lyrical Ballads" and I read first one and then the whole book. It conveyed to me no new doctrine, and yet the change it wrought in me could only be compared with that which is said to have been wrought on Paul himself by the Divine apparition. Looking over the "Lyrical Ballads" again, as I have looked over it a dozen times since then, I can hardly see what it was which stirred me so powerfully, nor do I believe that it communicated much to me which could be put in words. But it excited a movement and a growth which went on till, by degrees, all the systems which enveloped me like a body gradually decayed from me and fell away into nothing....
>
> There is, of course, a definite explanation to be given of one effect produced by the "Lyrical Ballads." God is nowhere formally

deposed, and Wordsworth would have been the last man to say that he had lost faith in the God of his fathers. But his real God is not the God of the Church, but the God of the hills, the abstraction Nature, and to this my reverence was transferred. Instead of an object of worship which was altogether artificial, remote, never coming into genuine contact with me, I had now one which I thought to be real, one in which literally I could live and move and have my being, an actual fact present before my eyes.[24]

This account of the unanticipated epiphany experienced by White would certainly appear to give unsolicited endorsement to the poet Percy Bysshe Shelley's conception of poets as "hierophants"—that is, priests.[25]

The Sage of the Lakes

Even in his lifetime Wordsworth came to be regarded as something of a guru. Many of the great and good in Victorian England, such as George Eliot and even the convinced religious skeptic Harriet Martineau, visited him in his remote home of Dove Cottage. When John Stuart Mill made his acquaintance with Wordsworth's work in 1828 he reported feeling free of the low-level depression that had long plagued him.[26] The influential critic and chaplain to Queen Victoria, Stopford Brooke, delivered a series of lectures in 1872 in which he praised Wordsworth's *personal* view of nature, contrasting it with Alexander Pope's soulless eighteenth-century conception: "But who is this person? Is she only the creation of imagination, having no substantive reality beyond the mind of Wordsworth? No, she is the poetic impersonation of an actual Being, the form which the poet gives to the living Spirit of God in the outward world."[27]

As F. W. H. Myers justly commented, Wordsworth's "exponents are not content to treat his poems on nature simply as graceful descriptive pieces, but speak of him in terms usually reserved for the originators of some great religious movement."[28] Myers made the further point that Wordsworth gave rise not just to a minority group of high-culture admirers but to a popular revolution in ordinary people's thinking: "Therefore it is that Wordsworth is venerated; because to so many men—indifferent, it may be, to literary or poetical effects, as such—he has shown by the subtle intensity of his own emotion how the contemplation of Nature can be made a revealing agency, like Love or Prayer—an opening, if indeed there be any opening, into the transcendent world."[29]

The fact that Wordsworth's *Guide to the Lakes* was a bestseller is in itself an indication of his popular appeal to all kinds and conditions of Britons. As Jonathan Bate puts it, "In Wordsworth's youth, genteel tourists came to the Lake District in search of the picturesque. After the advent of the railway, the urban middle and working classes came in search of Wordsworth."[30]

Disciples at Home and Abroad

Such was the conception of Wordsworth which was presently to radiate outwards from the British Isles. In 1848 Ralph Waldo Emerson is on record as having paid a visit to the then-aged Wordsworth. Emerson would go on to shape the thinking of Henry David Thoreau, who developed a distinctly American strain of Romantic retreat in the woods of Walden. The American philosopher William James was an admirer of Myers's exegesis of Wordsworth, and like Mill, reported being "restored to sanity" by his reading of the poet.[31] Impatient of the old three-tier universe (heaven, earth, hell in descending order) traditional to an older Christian cosmogony, James saw the religious instincts of mankind as being situated largely in the subconscious, its apprehensions mediated by the "still small voice" independently of any particular belief system.

Overall, Wordsworth's achievement might be described as that of having given currency to the belief that religious feeling had its origins in a subjective apprehension of the sacred. Wrote Matthew Arnold: "More and more mankind will discover we have to turn to poetry to interpret life for us, to console us, to sustain us. Without poetry, our science will appear incomplete; and most of what now passes with us for religion and philosophy will be replaced by poetry.... For finely and truly does Wordsworth call poetry "the breath and finer spirit of all knowledge."[32]

Thus poetry challenges the Bible as sole authority in mediating truth. Turning orthodoxy on its head, a reading of Wordsworth encouraged the idea that the genesis of religion lay in essentially mystical apprehension. Such was the conception which underlay two works previously mentioned, William James's *Varieties of Religious Experience* (1902) and German theologian Rudolf Otto's *Das Heilige / The Idea of the Holy* (1917).[33] Otto, for instance, defined God as an awe-inducing power, something *ganz andere* (wholly different) from normal categories of human experience—a presence which can be apprehended but not defined in precise conceptual terms.

The Romanian-American scholar of world religions, Mircea Eliade, would later endorse this understanding of God. For Eliade religion is a primal experience: "It is not a matter of theoretical speculation, but of a primary religious experience that precedes all reflection on the world."[34] Eliade invoked the concept of cosmic sacrality, the idea that Nature in its entirety is little less than a "hierophany," a revelation of the sacred. Wordsworth undoubtedly played a seminal role in the evolution of that conception in European/North American thinking, since his poetry has for many had the irresistible effect of re-sacralizing what was by his own, post-Enlightenment day becoming a desacralized world.

The upshot was that, as Jonathan Bate put it, "thirty years after his death, the poet from an obscure nook of northern England, who in the first half of his life was mercilessly derided by the critics, was being compared to Jesus Christ."[35] Various theories have been advanced for that apotheosis. Fellow poet John Keats contended that Wordsworth surpassed even John Milton, since Milton in his *Paradise Lost* had essentially poetized the accepted biblical grand narrative whereas Wordsworth had searched his heart to produce fresh and more authentic religious insights.[36] More recently, world religions authority Karen Armstrong made the bold claim that Wordsworth succeeded in bringing his readers closer to God. In her recently published *Sacred Nature* she contended that "since the fourteenth century, we have built an entirely different notion of the sacred. By rationalizing nature and confining God to the heavens, we have so drastically reduced the divine that for many it has become either imperceptible or incredible."[37] And by contrast, Wordsworth's evocation of God-in-Nature relocated Him within the natural world.

Such was certainly the apprehension of William Hale White who, as with so many others, made recourse to Wordsworth to supplement and strengthen his faith against the backdrop of older verities being undermined. Wordsworth's promulgation of Nature as a divine disclosure was clearly felt by many in Britain and elsewhere to provide support and consolation in the wake of the widespread undermining of the officially accredited support provided by the Bible.

Sages in Dispute

When Charles Darwin came to publish his *Origin of Species* nine years after Wordsworth's death, the British public's understanding of the natural world had already had ample time to absorb the poet's influence. His was in essence a rival philosophy of nature which offered a counterweight to Darwin's vision of a godless nature evolving by fixed, mechanical laws. Paradoxically, for many people Wordsworth's theology of nature formed a more effective counterforce to Darwin's ideas than biblical orthodoxy itself since that orthodoxy had been undermined by the biblical Higher Criticism. The Wordsworthian vision, in contrast, was invulnerable to the kind of attacks suffered by orthodoxy since the poet did not have recourse to miraculous or mythological elements or claim them as empirical fact. The Wordsworthian worldview could not be disproved by cold-eyed Germanic logic, wrote Bate, for "in the post-Lyell, post-Darwin world, God was not dead, as Nietzsche had proclaimed, but rather buried within the deep strata of the human psyche—the territory that Wordsworth had excavated in *The Prelude*."[38]

Alvar Ellegard found considerable resistance to Darwin on a number of purely logical grounds. For instance, the transmutational, descent-with-modification theory was rejected since, as Darwin himself was obliged to admit, there was no fossil evidence to support his conjecture. People were also unwilling to accept the enormous role of chance postulated for evolutionary processes, so that for the later

Victorian writer Samuel Butler, proponents of Darwinism were essentially "apostles of luck" with little statistical credibility on their side.[39] Yet aside from logic there was another reason for people's more positive affective response to Wordsworth when contrasted with their reactions to Darwin, namely, the moral stature of the two men's different philosophies. Darwin's philosophy of natural selection derived from his reading of Thomas Malthus and the demographer's survival-of-the-fittest/devil-take-the-hindmost ideas. Not surprisingly, such a brutal philosophy won Malthus many enemies on the Left, and it is thought that Charles Dickens chose "Population Malthus" as a basis for his deeply unappealing character of Scrooge in order that the iconic curmudgeon might act as an implicit reproach to a Malthusian philosophy deemed ignoble and inhumane.

No greater contrast to Malthus/Darwin could be found than Wordsworth, with his radically different attitude to social issues. Many of his poems, such as "Michael" and "The Old Pedlar" derive from conversations Wordsworth initiated with afflicted, dispossessed, or otherwise unregarded members of society whom he encountered on his Lakeland wanderings, persons whom most people of Wordsworth's social station might have elected to pass by on the other side. For the poet, these interlocutors possessed just as much dignity and worth as their ostensible betters. In that sense, Wordsworth's poems might be understood as a fitting accompaniment in their political orientation to the novels of Charles Dickens. Wordsworth's poetry shows the outworking of an internalized, practical Christianity.

As Robert Ryan observed, "His poetry was found to be of continuing service in the maintenance of religious belief, even if the faith was so attenuated as no longer to resemble [orthodox] Christianity."[40] That contention seems to me to be a fair one, and one might add to it that the poet has played some considerable part in the fact that his native land, often now labeled a post-Christian country, remains, contrary to the spirit of Darwinian nihilism, far from being a land without spiritual convictions. Indeed, among many of those not content to turn a blind eye to the implications of modern science, there is a growing awareness of less-evident dimensions of reality. The traditional methodology used by the legatees of Enlightenment thinking is increasingly revealing its limitations, bringing about a shift in people's thinking, a shift to which we turn in the following chapter.

7. Darwinism and the Loss of the Enlightenment Paradigm

L ONG BEFORE THE AGE OF DARWIN, THE SUBJECT OF EVOLUTION had been viewed as intimately linked to politics and religion. Those who believed in transmutation of species tended to be on the side of the religious doubters, and on the left politically.[1] It was commonly supposed that if you believed in an autonomous, self-fashioning and self-sustaining Nature, there would arise a corresponding impulse to define yourself as the independent product of natural processes rather than as the designed creation of a supernatural originator.

If then, as figures as various as Sigmund Freud and Jacques Monod have claimed, belief in the divine was a response to human feelings of insecurity and contingency, then that revolutionary declaration of independence from the divine we term atheism must count as a response to the countervailing human desire for untrammeled personal freedom. That aspiration had its political correlative in the desire to rise up against the forces of overbearing lordship (secular or divine) and forge one's own destiny. Such was the guiding spirit of the age in which Erasmus Darwin grew to maturity and which became the crucible of so many revolutionary impulses. As M. H. Abrams explained, by dint of an "assimilation of biblical and theological elements to secular frames of reference,"[2] hope for radical transformation was, in Enlightenment thinking, seen as achievable in historical time (as opposed to the distant, on-the-stroke-of-Doomsday transformation emphasized by Christian apocalyptic). The practical effect of this sea change from divine to secular was to translate scriptural prophecy rather freely into pre-emptive revolutionary action—"people doing it for themselves" to improve their condition by the form of direct action that did away with the need for apocalyptic hopes.

As the examples of the American and French revolutions showed, this could involve bloodshed, yet the pious hope was that the violence would have a cleansing effect which would ultimately lead to the greater good. As Roy Porter observed, it was Erasmus Darwin whose "man-centered view of man making himself" resulted in his "Promethean vision of infinite possibilities. God had become a distant cause of causes; what counted was man acting in Nature. The theodicy, the master narrative, had become secularized."[3] His was a new, secular cosmogony in explicit opposition to the older European master-narrative enshrined in the Bible and the twelve books of John Milton's *Paradise Lost*.

Erasmus was a quintessential legatee of Enlightenment pre-possessions. The thinkers of the Enlightenment wished to distance themselves from anything that smacked of religious "superstition." This led to the determination to assert a unilateral declaration of independence from the metaphysical sphere in favor of purely "scientific" modes of explanation. But the last century of decidedly counter-intuitive scientific discoveries has prompted many to acknowledge that the Enlightenment dream of encompassing the whole of reality in some grand materialist theory of everything was overly ambitious, even hubristic. The project, in short, has been forced into a reluctant retreat.[4]

The seventeenth-century discoveries of Sir Isaac Newton were in the course of Western intellectual history to become distilled into a cosmic worldview that set the pattern for every area of intellectual endeavor; but more recently we have had considerable reason to doubt whether that mechanistic framework tells the whole story.

The Newtonian understanding rested on the confident conception of the universe as a great machine acting according to predictable laws. The deeply religious Newton, to be clear, was not an atheist, and in affirming that "in him [God] are all things contained and moved,"[5] he was clearly a religious theist rather than a deist. Nevertheless, Newton's laws and his simplifying methods for reaching his discoveries strongly influenced later scientists toward a "Newtonian" way of approaching the world, always with the hope of mastering it mathematically and, by extension, mechanically.

In 1838 Charles Babbage, the Cambridge mathematician and originator (with Lord Byron's daughter, Ada Lovelace) of the first rudimentary computer, claimed that all mysteries of the natural world could be more credibly explained by natural laws than by supernatural acts. That thesis, as observed above, had been successfully tested by Charles Lyell in his demonstration that currently observable causes, acting at modest rates throughout the immensity of geological time, could build the full panoply of earthly events, from seas to mountains. According to the new doctrine of geological uniformitarianism, there was no longer any need for a divine hand to push up mountain ranges overnight. Once the Newtonian paradigm in cosmology had won acceptance in other branches of scientific enquiry, there followed a predictable amount of follow-my-lead behavior as it came to be thought that all scientific explanations should henceforth remain congruent with that paradigm. Rightly or wrongly, Newtonianism had become not just a set of mathematical laws in the discipline of physics but a habit of mind for approaching the world.

It was within that context of understanding that Charles Darwin's conjectures about the mysterious processes of speciation were integrated into that great, overarching meta-narrative of the age, which sought to reduce all things to natural causes—or in the case of the more modest project of *methodological* naturalism, which even many Christian scientists adhered to, to restrict scientific hypotheses to purely naturalistic causes and to see how far such a project could take them. Evolution had hitherto been viewed as a pseudo-science, but the postulation of the chance variation/natural selection mechanism as its "true cause," along with the theory's adherence to the increasingly influential rule of methodological naturalism, allowed it to shelter under the aegis of the scientific meta-narrative and so find a place in the hallowed domain of science proper.

Did not Darwinian theory dovetail satisfyingly with those other naturalistic approaches to life and the universe which had been gathering momentum in the late eighteenth and early nineteenth centuries? Well then, the prevailing attitude seemed to be, it is on the side of history, so let us welcome it and not be overly concerned with its apparent shortcomings. This attitude appeared to make perfect sense in the period circa 1860–1930, but scientific developments in the late 1920s and mid-1940s came together to challenge naturalized Newtonian assumptions of *what reality really is.* As a corollary, we have been prompted to seriously question what place Darwinian theory has within an emerging scientific paradigm that has fatally undermined older scientific certainties together with that cozy nineteenth-century consensus in which Darwinism had been accorded such a comfortable billet.

A New Paradigm

A century after the Enlightenment, in the course of the Industrial Revolution, science came to command allegiance as an unstoppable source of knowledge, viewed by increasing numbers as a power capable of emancipating humankind from its dependence on the widely questioned claims of organized religion. In a very qualified sense that proud boast had some force to it. During the preceding hundred years, scientists and engineers had been extraordinarily successful in understanding some natural phenomena and controlling the destructive force of many others. The flood which caused death and destruction on both sides of the Bristol Channel in 1607, and similar tragedies, were no longer understood as the working out of God's wrath upon sinful humanity, as was frequently intoned in contemporary pamphlets in flood-stricken South Wales and the West Country in the early seventeenth century. When Canvey Island on the south coast of England was flooded in 1953, precious few would have claimed that event as evidence of divine wrath. Rather were people minded to take practical precautions against any further storm surges by building high embankments against encroachments of the sea. Such an understanding was to undergird the construction of that wonder of modern hydraulic engineering, the Thames Barrage, so successful in preventing major flooding of the capital in 2007.

There is, however, all the difference in the world between such pragmatic, human-engineering interventions as flood barriers and the ability to understand the *mainsprings* of the natural world.

That age-old mystery is considerably more complex than Darwin and his contemporaries assumed. Some of us who imagined that we had escaped Plato's cave and at last seen things as they truly are, now realize that this itself was an illusion and that we remain that band of prisoners with no direct knowledge of the deepest workings of nature. Our ignorance of what Lucretius termed the true nature of things has been unsparingly revealed by the work of pioneering twentieth-century scientists such as Max Planck, Albert Einstein, Niels Bohr, Werner Heisenberg, Carlo Rovelli, and others. Their work has left deep fissures in the Newtonian/Enlightenment paradigm. Werner Heisenberg, who introduced the famous Uncertainty Principle to an astonished world in 1927, was the first to demonstrate the fact that the regularities of the larger Newtonian universe simply did not apply to the subatomic world. Scientists can no longer deliver certainty and predictability in the aftermath of such disorienting advances in physics.

As numerous science media stories have soberly explained, it is not possible in the world of the very small to measure the position of an object and its momentum at one and the same time. Newtonian logic clearly has a strictly limited applicability in a realm where only approximate knowledge can be achieved and where the new watch-word of probabilism reigns supreme. Mysteries arrived on the heel of mysteries. Photons, as it turns out, cannot be rightly characterized as either waves or particles, but instead must be regarded as some baffling hybrid of the two. Einstein's theory of general relativity was a great triumph, taking the Newtonian model a level deeper into reality; but the highly successful Einsteinian model, as it turns out, is not compatible with quantum mechanics. Their incompatibility suggests some deeper law as yet undiscovered, but the search for any such law remains not just elusive but marked by what some have come close to characterizing as a comedy of errors. Sabine Hossenfelder captured it well in the title of her 2018 book, *Lost in Math*.

Even quantum physics itself, content to rest modestly in the probabilistic, finds itself baffled by the deepest layers of the subatomic realm, as physicists struggle to explain the results of, for example, the now-famous double slit experiment, for which there are at least a half dozen competing interpretations. (To give one a quick taste for how far physicists have reached in their attempts to grope their way out of the dark, one of those interpretations is known as the "many-worlds interpretation.")

After such bewildering advances in quantum mechanics, with its (only) probabilistic laws, many insisted it was now necessary to revisit the subject of causality and even reality itself. Early twentieth-century British scientist Sir Arthur Eddington, for instance, claimed that religion became eminently possible for a reasonable scientific person in the year 1927. That was the year of the promulgation of the Uncertainty Principle, which essentially announced that "all bets are off" on mankind's aspirations to fully understand nature. We were, Eddington strongly hinted, dependent on powers far beyond our present comprehension. Or as geneticist J. B. S. Haldane put, "The universe is not only queerer than we suppose, but queerer than we *can* suppose."[6]

Biology, as well, in delving deeper and deeper into the cellular realm, has uncovered layers of intricacy that have exploded our facile descriptions of biological structures being akin to man-made machines. Whether one regards the engineer as God or some wholly natural process stretched over eons of time, it is now glaringly apparent that this engineer far outstrips our best human engineers, having conjured biological marvels as yet far beyond anything we can understand, much less build. Biologists find before them a previously unsuspected subcontinent of ultra-diminutive reality, one that has only revealed its bare existence in the last seven decades or so, thanks to the invention of such tools as the electron microscope.

Such scientific findings, with their negative implications for humanity's erstwhile Promethean conviction of its coming omniscience, have delivered a staggering blow to the Enlightenment meta-narrative. Against its vision of an all-knowing science just over the next horizon, our modern and postmodern world has brought with it a grudging acknowledgment that the goal of comprehensive scientific certainty seems to recede rather than draw nearer with each new fundamental discovery. The notion that just around the corner is a theory of everything no longer commands the unqualified respect it once did, and notably it is science itself (rather than modish intellectual fashion) that has called for this new sobriety of mood. In the quest to explain perennial mysteries, it has been justly observed that "in the final analysis, absolute materialism does not triumph because it cannot fully explain the nature of reality."[7]

We now find ourselves forced to acknowledge that the picture of the world we gain from classical Newtonian physics has so little in common with what quantum physics tells us is the underlying reality of things that they could be describing two different universes rather than the same one. Although materialists have striven mightily to square the quantum domain with their worldview, quantum physics seems to have provided far more aid and comfort to dualistic and idealistic thinkers, as they contemplate a Lilliputian realm characterized by immaterial mathematics, uncertainty, and the mysterious effects of observing minds. The evasive habit of mind which would separate off the quantum world from direct consideration as if it were a thing apart is, according to *Spiritual Science* author Steve Taylor, a primary reason for the limited impact that the findings of quantum physics have had on the prevailing materialist worldview.[8] Reality, in other words, is being cherry-picked to support an archaic world picture, as if quantum physics had never been discovered and, it might be added, as if there existed no terminal impasse about understanding the brain and consciousness or other insoluble mysteries of human physiology and neurology.

Undoubtedly Newtonianism, particularly its modern, naturalistic instantiation, is more comforting than the quantum realm because it gives us the illusion we have a handle on the world, and hence a modicum of control over it; but in this it nourishes a false hope, since, as Christian Bandea puts it, science is manifestly unable to "peer across the existential brink, into the realm of pure creation."[9]

Whose Reality?

Whether we like it or not, Erasmus Darwin's simple and predictable world is no more, and we now find ourselves subject to a profoundly mysterious cosmos. It appears to me that the only intellectually defensible position to adopt in the light of such unanticipated scientific advances is the modest one of keeping an open mind. By contrast, that group of twenty-first century thinkers dubbed the New Atheists[10] have clung anachronistically to the same would-be omniscient paradigm of reality as that in which Erasmus Darwin reposed his faith. But whereas Erasmus had the excuse of knowing nothing of the deeper reaches of physical reality into which modern scientific advances have given us at least fleeting glimpses, the same excuse cannot be pleaded for the New Atheists.

Under the illusion of being the "brights" (their term)—that is, the enlightened ones—the New Atheists appear instead to have become the doctrinaire victims of a peculiarly modern form of obscurantism, doggedly clinging to an obsolete worldview while ignoring the implications of much cutting-edge science. Their outlook has little in common with that of Charles Darwin, whose later years were marked by what Peter Vorzimmer terms "frustrated confusion."[11] In that respect, it might be added, he should be posthumously welcomed as a prefiguration of postmodern man in that his unresolved tensions anticipated the decidedly non-omniscient spirit of our present age.

The New Atheists' exhibitionistic unbelief, by contrast, seems more akin to that of Charles's overconfident grandfather. Erasmus was certainly more prone to gratuitous displays of atheistical dissent than was his grandson. This was clearly shown in a minor controversy in his hometown of Lichfield, arising from the provocative inscription he once had embossed on the exterior of his coach, the offending words being *E Conchis Omnia* (everything comes from seashells). One is reminded of Richard Dawkins paying to have an atheistic message emblazoned on the sides of London buses, or of his confidently reducing biological reality to the hopelessly reductionist figure of the "selfish gene."

As the later course of Darwin's scientific career demonstrated, his preferred way was the quietist one of avoiding conflict and controversy, made manifest in his dedication of the latter decades of his life to uncontroversial subjects such as earthworms. Yet Darwin's temperamental desire for an uncontroversial life tells only part of the story. The more substantive reason for his disinclination to join the ranks of Charles Bradlaugh, Annie Besant, and other materialist proselytizers of his day was that his latter years were accompanied by the grace to disavow previous claims to overweening certainty. At that stage of his life he felt duty-bound to acknowledge that he was not *completely* convinced of his own theory and that behind natural selection, he obscurely suspected, might lie some teleology hidden from humankind (see Chapters 2 and 5).

The bright new dawn of Erasmus Darwin's Enlightenment world has been well and truly replaced by the unsettling specter of "quantum weirdness." Indeterminacy and probabilism have emerged to subvert the Enlightenment conception of a predictable clockwork universe. In particular, we have been forced to acknowledge that the knowable and observable part of nature rests on, and is sustained by, invisible trestles of substrate reality which we are unable to fathom and to which Cartesian notions of predictability and comprehensibility do not apply. Wordsworth's intimation that there was something "more deeply interfused" in Nature than mankind could observe appears to have received a deal of confirmation from modern science.

Thus it is that we find a great mystery before us: Why should a planet of modest size, what astronomer Carl Sagan characterized as but a "pale blue dot" amid the vast reaches of interstellar space, have developed not just life but a wildly various and pulsing biosphere, one in such stark contradistinction to the sterile mortuary kingdom surrounding it? There is after all no point of comparison between the living and breathing Earth cradled in its proverbial Goldilocks zone and the endless deadness of the surrounding universe. We represent a cosmic exception so unusual as to require an altogether separate form of explanation from that employed to explain the inanimate universe. Indeed, the absolute disparity in existential status between our living planet and its surroundings prompts the inference that life—and in particular, sentient life—could not have developed without some form of foresight and an accompanying instrumental power to realize some originary vision intent on moving Earth beyond the otherwise unwavering rule of sterility.

There cannot even be any sensible talk of an inference to the best explanation when there are no helpful indices pointing to a purely naturalistic understanding aside from vague and desperate thought experiments about a "multiverse," conjectures that are not only untethered from observation and the physical reality of our universe but also lead to an erasure of the ground for trusting fundamental tenets of reason.[12] Hence, to account for the underlying nature of things by postulating the operation of chance in any of its claimed aspects (e.g., natural selection operating on random variation, jostling atoms miraculously cohering into the first self-reproducing biological entity, our universe winning a multiverse lottery of infinite universes) can represent only a wishing away of the unsolved mystery.

This applies as much to the present day as it does to Epicurus and Lucretius some two millennia ago. Leading astronomers such as Fred Hoyle, Paul Davies, and Charles Townes have emphasized that the benign cosmic arrangements we enjoy could hardly have arisen by chance, and have the appearance of an intentional arrangement. Townes was quite explicit. "Intelligent design, as one sees it from a scientific point of view, seems to be quite real," he commented. "This is a very special universe: it's remarkable that it came out just this way. If the laws of physics weren't just the way they are, we couldn't be here at all. The sun couldn't be there, the laws of gravity and nuclear laws and magnetic theory, quantum mechanics, and so on have to be just the way they are for us to be here."[13] Davies added that it is an ancillary consideration as to whether one conceives of the shaping force behind this providential arrangement as being the Christian God or some other unseen power, for *some* agency there must have been.[14]

The bottom line is that it is logically impossible to conceive of our planet as an arbitrary and accidental collocation of atoms, objects, and life-forms—as both ancient Lucretianism and Lucretianism's modern legatee, present-day orthodoxy, insist in the teeth of a mountain of counter-indications. Even if we are obliged to concede that the ultimate seat of authority cannot be apprehended in anthropomorphic categories of understanding, a basic respect for the balance of probabilities should dictate that the existence of such an agency, albeit unfathomable, should at least be taken seriously in our current conversations.

At the end of the day there seems little exaggeration in asserting that advances in our understanding of everything from the very small (cells and the quantum realm) to the very large (the Big Bang and the fine-tuning of the laws and constants of nature for life), together with a growing appreciation for nature's as yet unplumbed layers of complexity, has led to an unexpected revival of the fortunes of Paley's once-derided natural theology.

Historically, the unfathomable subtleties of our terrestrial environment have been viewed as empirical markers for design. Such, as noted above, was the conception defended by various distinguished thinkers of the ancient world, including Plato, Aristotle, and Cicero, whose conclusions informed European thinking right up to Darwin's century, when Paley famously devoted a whole volume to the subject of design. Cicero's striking analogy of the sundial anticipated Paley's famous watch-on-the-heath analogy[15] by two millennia, and it is likely the classically educated Paley was influenced by the Ciceronian analogy. Cicero posed the matter thus:

> When you look at a picture or a statue you recognize that it is a work of art. When you follow from afar the course of a ship, you do not question that its movement is guided by a skilled intelligence. When you see a sundial or a water-clock, you see that it tells the time by design and not by chance. How then can you imagine that the universe as a whole is devoid of purpose and intelligence when it embraces everything, including these artefacts themselves and their artificers?[16]

The idea of intelligent design is just as much a commonsense, empirically based scientific observation as it is a formal philosophical theory or religious tenet, being as clear to most ancient Greeks and Romans as it is two millennia later to most citizens of the modern world. It seems intuitively right without benefit of formal elaboration of the philosophical kind and requires no commitment to a particular creed to make it intellectually compelling.

Darwin hoped to provide a totalizing explanation for the origin of life's great diversity, but more than 160 years after *The Origin of Species* arrived like a thunderclap, we find ourselves far from understanding even life's present operation, much less how precisely it all came to be millions and billions of years ago. Lyell's counsel of modesty toward matters biological now appears prescient. The last half century or so of microbiological research has shown us that Nature's generative programs are constructed in such a subtly coded web of biological imperatives that they as yet remain beyond human fathoming. They are, unquestionably, orders of magnitude beyond the most advanced creations of the most skilled of human artificers.

The habit of Darwin's later adherents to whistle in the dark while telling the rest of us "nothing to see here, job done, hurry along" only heightens my own awareness of what the early twentieth-century German theologian Rudolf Otto termed the "numinous" dimension of reality: a reality which can be sensed but not understood in precise conceptual terms.

There is surely considerable irony in the fact that Darwinism, when subjected to an unblinking rationalist critique, reveals itself to be so completely unconvincing as to propel a (historical) secularist like myself in the direction of theistically oriented meditations on life's ultimate realities.

ENDNOTES

PROLOGUE

1. This large issue is also dealt with by R. C. Sproul and Keith Mathison in *Not a Chance: God, Science and the Revolt against Reason*, 2nd ed. (Grand Rapids, MI: Baker, 2014).

2. Karl Marx to Friedrich Engels, December 19, 1860, *Marx & Engels Collected Works, Volume 41: Letters 1860–64* (London: Lawrence & Wishart, 2010), 232, https://www.hekmatist.com/Marx%20Engles/Marx%20&%20Engels%20Collected%20Works%20Volume%2041_%20Ka%20-%20Karl%20Marx.pdf .chrome-extension://efaidnbmnnnibpcajpcglclefindmkaj/https://www.hekmatist .com/Marx%20Engles/Marx%20&%20Engels%20Collected%20Works%20 Volume%2041_%20Ka%20-%20Karl%20Marx.pdf.

3. Karl Marx to Ferdinand Lassalle, January 16, 1861, *Marx & Engels Collected Works, Volume 41*, 245, http://hiaw.org/defcon6/works/1861/letters/61_01_16.html.

4. Friedrich Engels, *Dialectics of Nature* [1925], transl. Clemens Dutt [1934] (Moscow: Progress Publishers, 1986), 195, chrome-extension: //efaidnbmnnnibpcajpcglclefindmkaj/https://www.marxists.org /archive/marx/works/1883/don/don.pdf.

CHAPTER 1: DARWINISM AND ITS DISCONTENTS

1. Charles Darwin, *On the Origin of Species, by Means of Natural Selection* (London: John Murray, 1859), 5.

2. Cynthia Eagle Russett, *Darwin in America: The Intellectual Response 1865–1912* (San Francisco, CA: Freeman, 1976), 6. Emphasis added.

3. Darwin, *On the Origin of Species*, 131.

4. Quoted by Charles Darwin to Charles Lyell, December 10, 1859, *Darwin Correspondence Project*, Letter no. 2575, University of Cambridge, https://www.darwinproject.ac.uk/letter/?docId=letters/DCP-LETT-2575.xml.

5. Charles Darwin and Alfred Russel Wallace, "On the Tendency of Species to Form Varieties; and on the Perpetuation of Varieties and Species by Natural Means of Selection," *Journal of the Proceedings of the Linnean Society*, August 1858 (papers read out on July 1, 1858). See the accompanying letter of Charles Lyell to the Linnean Society, June 30, 1958, *Darwin Correspondence Project*, Letter no. 2299, University of Cambridge, https://www.darwinproject.ac.uk/letter?docId =letters/DCP-LETT-2299.xml.

6. Charles Darwin, *The Autobiography of Charles Darwin*, ed. Nora Barlow (New York: Norton, 1958), 122.

7. Samuel Haughton put the matter thus: "This speculation of Messrs. Darwin and Wallace would not be worthy of notice, were it not for the weight of authority of the names under whose auspices it has been brought forward [Lyell and Hooker]. If it means what it says, it is a truism; if it means anything more, it is contrary to fact." *Annual Address Delivered Before The Geological Society of Dublin*, February 8, 1859 (Dublin: M. H. Gill, 1859), 18, https://play.google.com/books/reader?id=dq4oAQAAMAAJ&pg=GBS.PA18&hl=en.

8. Samuel Haughton, "Biogenesis" [review of *Origin* by Charles Darwin], *Natural History Review* 7 (1860): 23–32, http://darwin-online.org.uk/converted/pdf/1860_Review_Origin_Biogenesis_Haughton_A1128.pdf.

9. Haughton, "Biogenesis."

10. Haughton, "Biogenesis."

11. Charles Darwin, *The Descent of Man and Selection in Relation to Sex* [1871], eds. James Moore and Adrian Desmond (London: Penguin, 2004).

12. St. George Jackson Mivart, *On the Genesis of Species*, 2nd ed. (New York: Macmillan & Co., 1871); see for example 145–153.

13. Mivart, *On the Genesis of Species*, 255. The work was framed as a studied response to Darwin's *Origin*.

14. David L. Hull, *Darwin and his Critics: The Reception of Darwin's Theory of Evolution by the Scientific Community* (Chicago: Chicago University Press, 1973) reproduces many of the more important reviews published within a decade of the appearance of the *Origin*.

15. See W. J. Dempster, *Evolutionary Concepts in the Nineteenth Century: Natural Selection and Patrick Matthew* (Edinburgh, UK: Pentland Press, 1996), 71.

16. See Alvar Ellegard, *Darwin and the General Reader: The Reception of Darwin's Theory of Evolution in the British Periodical Press 1859–1872* (Gothenburg, Sweden: Elanders, 1958).

17. Curtis Johnson, *Darwin's Dice: The Idea of Chance in the Thought of Charles Darwin* (Oxford, UK: Oxford University Press, 2015), 89–114 and passim.

18. Johnson, *Darwin's Dice*, xvii.

19. Alfred Russel Wallace to Charles Darwin, July 2, 1866, *Darwin Correspondence Project*, Letter no. 5140, University of Cambridge, https://www.darwinproject.ac.uk/letter/?docId=letters/DCP-LETT-5140.xml.

20. Darwin notably uses the architect metaphor in the *Variation of Animals and Plants under Domestication* (London: John Murray, 1968), vol. 2, chap. 30, 248–249. Lyell's rejection of the idea is spelled out in a letter to Darwin, June 15, 1860, *Darwin Correspondence Project*, Letter no. 2832A, University of Cambridge, https://www.darwinproject.ac.uk/letter/?docId=letters/DCP-LETT-2832A. See also Darwin to Lyell, June 17, 1860, *Darwin Correspondence Project*, Letter no. 2833, University of Cambridge, https://www.darwinproject.ac.uk/letter/?docId=letters/DCP-LETT-2833.xml. Johnson, *Darwin's Dice*, 137, discusses in detail this metaphor and its weakness.

21. See further discussion of this point in Johnson's *Darwin's Dice*, 136–143.

22. Johnson, *Darwin's Dice*, 227.

23. Jessica Riskin, *The Restless Clock: A History of the Centuries-Old Argument over what Makes Us Tick* (Chicago: University of Chicago Press, 2016). See chap. 6, "Dilemmas of a Self-Organising Machine," 189–213, and chap. 7 ,"Darwin between the Machines," 213–249.

24. William Paley, *Natural Theology: Or Evidences of the Existence and Attributes of the Deity, Collected from the Appearances of Nature [1802]*, eds. Matthew D. Eddy and David Knight (Oxford, UK: Oxford University Press, 2008), 7–8.

25. Riskin, *Restless Clock*, 363.

26. Riskin, *Restless Clock*, 4.

27. Riskin, *Restless Clock*, 4.

28. Darwin would not be deterred by this intuitive sense that construction of the eye required more than mere mindless processes. As he explained to Asa Gray in a February 1860 letter, "About weak points I agree. The eye to this day gives me a cold shudder, but when I think of the fine known gradations, my reason tells me I ought to conquer the cold shudder." February 8 or 9, 1860, *Darwin Correspondence Project*, Letter no. 2701, University of Cambridge, https://www.darwinproject .ac.uk/letter/?docId=letters/DCP-LETT-2701.xml#DCP-BIBL-6014.

29. Daniel Dennett, *Darwin's Dangerous Idea: Evolution and the Meanings of Life* (London: Allen Lane, 1995), 59.

30. Richard Dawkins, *The Blind Watchmaker: Why the Evidence of Evolution Reveals a Universe Without Design* (London: Penguin, 1986), 159–160.

31. Scott Hahn and Benjamin Wiker, *Answering the New Atheism: Dismantling Dawkins' Case Against God* (Steubenville, OH: Emmaus Road, 2008), 11, 13.

32. Richard Dawkins once pronounced with seigneurial condescension: "Pretend as they will to scientific credentials, the anti-evolution propagandists are always religiously motivated, even if they try to buy credibility by concealing the fact. In most cases, they know deep down what to believe because their parents recommended an ancient book that tells them what to believe. If the scientific evidence learned in adulthood contradicts the book, there must be something wrong with the scientific evidence. Since all radiometric dating methods agree that the earth is thousands of millions of years old, something obviously has to be wrong with all radiometric dating methods. The holy book of childhood cannot be, must not be, wrong." Dawkins, *The Blind Watchmaker*, xiv–xv.

33. Bill Gates, Nathan Myhrvold, and Peter Rinearson, *The Road Ahead*, rev. ed. (Harmondsworth, UK: Viking Penguin Group, 1996), 228.

34. Eric H. Anderson, "A Factory That Builds Factories That Build Factories That…," in Thomas Y. Lo et al., *Evolution and Intelligent Design in a Nutshell* (Seattle, WA: Discovery Institute Press, 2020), 69.

35. Michael J. Behe, *Darwin's Black Box: The Biochemical Challenge to Evolution* (New York: The Free Press, 1996) and *A Mousetrap for Darwin: Michael J. Behe Answers His Critics* (Seattle, WA: Discovery Institute Press, 2020).

36. For numerous cases involving the human body, see Steve Laufmann and Howard Glicksman, *Your Designed Body* (Seattle, WA: Discovery Institute Press, 2022).

37. Thomas Huxley, *Darwiniana: Essays* (London: Macmillan, 1893), 20, https:// archive.org/details/b21500009/page/20/mode/2up?q=%22enable+naturalists%22.

38. Eugene Koonin, "The Biological Big Bang Model for the Major Transitions in Evolution," *Biology Direct* 2 (2007), article 21.

39. Koonin, "The Biological Big Bang Model."

40. Marcel-Paul Schützenberger, "The Miracles of Darwinism," *Origins & Design* 17, no. 2 (November 1996), https://www.arn.org/docs/odesign/od172/schutz172.htm. This interview originally appeared in French in *La Recherche* (January 1996).

41. See the early review of the *Origin* by Fleeming Jenkin (1867) reprinted with discussion in *Darwin and His Critics: The Reception of Darwin's Theory of Evolution by the Scientific Community*, ed. David L. Hull (Chicago: Chicago University Press, 1973), 302–350. Jenkin scorned the idea that even a veritable infinity of time could produce large changes (Jenkin was the distinguished Scottish academic who worked with Lord Kelvin to secure the transatlantic cable).

42. John Milton, *Paradise Lost*:
    The Earth obey'd, and straight,
    Op'ning her fertile womb, teem'd at a birth
    Innumerous living creatures, perfect forms,
    Limb'd and full grown.

43. "New Trends in Evolutionary Biology: Biological, Philosophical and Social Science Perspectives," Royal Society of London, November 7–9, 2016, https://royalsociety.org/pastevent/?requestedPath=%2fscience-events-and-lectures%2f2016%2f11%2fevolutionary-biology%2f. See also, "Why the Royal Society Meeting Mattered, in a Nutshell," *Evolution News and Science Today*, December 5, 2016, https://evolutionnews.org/2016/12/why_the_royal_s/.

44. Charles Darwin, *The Origin of Species by Means of Natural Selection; Or, the Preservation of Favored Races in the Struggle for Life*, 6th ed. (London: Murray, 1872), 429.

45. See on this point Alfred Kelly, *The Descent of Darwin: The Popularization of Darwinism in Germany 1860–1914* (Chapel Hill, NC: University of North Carolina Press, 1981), 33.

46. Charles Darwin to James Hooker, February 1, 1871, *Darwin Correspondence Project*, Letter no. 7471, University of Cambridge, https://darwinproject.ac.uk/letter/?docId=letters/DCP-LETT-7471.xml.

47. Inside human cells, coded messages in the DNA are translated by RNA into some twenty distinct amino acids. This amino-acid alphabet in turn is used to code for the working molecules of protein, which carry out a wide array of life-essential tasks.

48. The 2016 meeting of the Royal Society of London, "New Trends in Evolutionary Biology," exemplified this burgeoning class of Darwin doubters from within the larger tent of mainstream biology. For an analysis of the event, see Paul Nelson and David Klinghoffer, "Scientists Confirm: Darwinism Is Broken," *CNSNews*, December 13, 2016, https://web.archive.org/web/20161216114652/http://www.cnsnews.com/commentary/david-klinghoffer/scientists-confirm-darwinism-broken.

CHAPTER 2: METAPHORS, ANALOGIES, SUBTEXTS

1. Charles Darwin, *On the Origin of Species by Means of Natural Selection; Or, the Preservation of Favoured Races in the Struggle for Life*, 2nd ed. (London: John Murray, 1860), 490.

2. These controversies typically involved "efforts to transform the Darwinian theory with a metaphysics of providence and progress which, by supplanting

causo-mechanical explanations, could secure a teleology and a theodicy on an evolutionary basis." James R. Moore, *The Post-Darwinian Controversies: A Study of the Protestant Struggle to Come to Terms with Darwinism in Britain and America 1870–1900* (Cambridge, UK: Cambridge University Press, 1981), 15.

3. Darwin, *On the Origin of Species*, 84.

4. Sander Gliboff, *H. G. Bronn, Ernst Haeckel and the Origins of German Darwinism: A Study in Translation and Transformation* (Cambridge, MA: MIT Press, 2008), 136.

5. Basil Willey, *Darwin and Butler: Two Versions of Evolution* (London: Chatto and Windus, 1960), 30. Willey refers here to the views expressed by Edward Bouverie Pusey in his sermon *Un-Science, Not Science, Adverse to Faith* (Oxford, UK: Devonport Society of the Holy Trinity, 1878), https://anglicanhistory.org/pusey/unscience.pdf.

6. Samuel Wilberforce, "On the Origin of Species," *Quarterly Review* (1860): 224–265. On Wilberforce's achievements see Standish Meacham, *Lord Bishop: Life of Samuel Wilberforce 1805–1873* (Cambridge, MA: Harvard University Press, 1974).

7. Dov Ospovat, *The Development of Darwin's Theory: Natural History, Natural Theology and Natural Selection 1839–1859* (Cambridge, UK: Cambridge University Press, 1981), especially 207–212. See also Steve Dilley's recent work on Darwin's use of theology in the *Origin*: "Charles Darwin's Use of Theology in *The Origin of Species*," *British Journal for the History of Science* 44, no. 1 (2012): 29–56.

8. See Robert J. Richards, *The Romantic Conception of Life: Science and Philosophy in the Age of Goethe* (Chicago: University of Chicago Press, 2002), especially 514–553, and Andrea Wulf, *The Invention of Nature, The Adventures of Alexander von Humboldt* (London: John Murray, 2015), especially 217–234.

9. Charles Darwin, "Darwin's Notebooks on Transmutation of Species: Part I. First Notebook [B] (July 18371838)," ed. Gavin de Beer, *Bulletin of the British Museum Historical Series* 2, no. 2 (January 1960): 53, https://darwin-online.org.uk/content/frameset?itemID=F1574a&keywords=creates%20creator&viewtype=text&pageseq=31. Darwin and Asa Gray discuss "designed laws." See for instance Charles Darwin to Asa Gray, May 22, 1860, *Darwin Correspondence Project*, University of Cambridge, Letter no. 2855, https://www.darwinproject.ac.uk/letter/?docId=letters/DCP-LETT-2814.xml; and Charles Darwin to Asa Gray, July 3, 1869, *Darwin Correspondence Project*, Letter no. 2855, University of Cambridge, https://www.darwinproject.ac.uk/letter/?docId=letters/DCP-LETT-2855.xml.

10. Ospovat, *The Development of Darwin's Theory*, 30.

11. The phrase is that of Neal C. Gillespie, *Charles Darwin and the Problem of Creation* (Chicago: University of Chicago Press, 1979).

12. A. O. J. Cockshut, *The Unbelievers: English Agnostic Thought 1840–1890* (London: Collins, 1964), 10.

13. *Charles Darwin: His Life Told in an Autobiographical Chapter and in a Selected Series of His Published Letters*, ed. Francis Darwin (London: John Murray, 1892), 31, https://www.gutenberg.org/files/38629/38629-h/38629-h.htm.

14. John Milton, *Paradise Lost* [1667], ed. Christopher Ricks (London: Penguin, 1989), Book I, lines 22–26.

15. Milton, *Paradise Lost*, Book I, lines 122–124.

16. Milton, *Paradise Lost*, Book II, lines 822–823.

17. William Empson, *Milton's God* (London: Chatto and Windus, 1965). On Satan as the true hero of the work, see especially 36–90.

18. The William Blake comment in full reads, "The reason Milton wrote in fetters when he wrote of Angels and God, and at liberty when of Devils and Hell, is because he was a true poet, and of the Devil's party without knowing it." William Blake, "The Voice of the Devil," in *The Marriage of Heaven and Hell* [1790–1793] (Boston: John W. Luce and Co., 1906), https://www.gutenberg.org/ebooks/45315. See also Margaret Kean, *John Milton's Paradise Lost: A Sourcebook* (London: Routledge, 2005), 51.

19. In his diary for October 7, 1660, Samuel Pepys wrote, "I perceive my Lord is grown a man very indifferent in all matters of Religion," https://www.pepysdiary.com/diary/1660/10/07/.

20. Claire Tomalin, *Samuel Pepys: The Unequalled Self* (London: Penguin, 2002), 114.

21. Charles Darwin to Asa Gray, May 22, 1860, *Darwin Correspondence Project*, Letter no. 2814, University of Cambridge, https://www.darwinproject.ac.uk/letter/DCP-LETT-2814.xml.

22. Charles Darwin to J. D. Hooker, February 1, 1871, *Darwin Correspondence Project*, Letter no. 7471, University of Cambridge, https://www.darwinproject.ac.uk/letter/DCP-LETT-7471.xml.

23. Charles Darwin to John Fordyce, May 7, 1879, *Darwin Correspondence Project*, Letter no. 12041, University of Cambridge, https://www.darwinproject.ac.uk/letter/?docId=letters/DCP-LETT-12041.xml.

24. Charles Darwin, *The Autobiography of Charles Darwin 1809–1882*, ed. Nora Barlow (New York: Norton, 1958), 92, https://archive.org/details/autobiographyofc0000nora/page/92/mode/2up?q=%22immense+and+wonderful%22.

25. Darwin, *The Autobiography of Charles Darwin*, 93.

26. Darwin, *The Autobiography of Charles Darwin*, 87.

27. Darwin, *The Autobiography of Charles Darwin*, 93–94.

28. Darwin, *The Autobiography of Charles Darwin*, 85.

29. Darwin, *The Autobiography of Charles Darwin*, 87.

30. Darwin, *The Autobiography of Charles Darwin*, 87.

31. Darwin, *The Autobiography of Charles Darwin*, 90.

32. See Tim Whitmarsh, *Battling the Gods: Atheism in the Ancient World* (London: Faber and Faber, 2016), 173–185.

33. Cited by Andrew S. Reynolds, *Understanding Metaphors in the Life Sciences* (Cambridge, UK: Cambridge University Press, 2022), 2.

34. See Maud Bodkin, *Archetypal Patterns in Poetry* [1927] (New York: Vintage, 1958).

35. Misia Landau, *Narratives of Human Evolution* (New Haven, CT: Yale University Press, 1993).

36. Reynolds, *Understanding Metaphors in the Life Sciences*, 89n59.

37. Charles Darwin, *Questions about the Breeding of Animals* [1839], ed. John van Wyhe (London: Stewart and Murray, 2002). For an online edition (with an introduction by Gavin de Beer), see https://darwin-online.org.uk/content/frameset?itemID=F263&viewtype=text&pageseq=1. Question 3 is revealing of the connection Darwin perceives between animal breeding and his theory of natural selection.

38. On the longevity of the semi-deified *Natura* concept, see Barbara Newman, *God and the Goddesses: Visions, Poetry and Belief in the Middle Ages* (Philadelphia: Pennsylvania University Press, 2003). Newman pointed out that the nineteenth century was the last great age of Nature's literary career, in which she was given a strong lease of life by many of the poets whom Darwin himself read, such as Wordsworth and Tennyson. There was even a quasi-Spinozan tendency among some of the Romantics, Wordsworth especially, to "merge Nature with Nature's God" (137).

39. An account of the supportive community enabling Darwin to concentrate on his writing is given by Edna Healey in her *Emma Darwin: The Inspirational Wife of a Genius* (London: Headline, 2001), esp. 230–247.

40. Gillian Beer, *Darwin's Plots: Evolutionary Narrative in Darwin, George Eliot and Nineteenth-Century Fiction* (London: Routledge, 1983), 6.

41. George Levine makes the fair point that Darwin "had no pretensions to high literary style and no concern to produce a book that would be beautiful or moving—as, in fact, it sometimes is. He might even have been alarmed if he were to have found himself in this book… juxtaposed to Dickens and George Eliot instead of to Cuvier, say, or Lyell." Levine, *Darwin the Writer* (Oxford, UK: Oxford University Press, 2011), 15–16.

42. Levine, *Darwin the Writer*, 88.

43. Stephen Jay Gould, *Bully for Brontosaurus* (New York: Norton, 1991), 264. Cited by Michael Ruse, *The Gaia Hypothesis: Science on a Pagan Planet* (Chicago: University of Chicago Press, 2013), 34.

44. "Natural Selection: The Trouble with Terminology, Part I," *Darwin Correspondence Project*, University of Cambridge, https://www.darwinproject.ac.uk/commentary/evolution/natural-selection-trouble-terminology-part-i.

45. Levine, *Darwin the Writer*, 87.

46. Richards, *The Romantic Conception of Life*, 537.

47. *Darwin the Writer*, 111n6.

48. Jerry Fodor and Massimo Piattelli-Palmarini, *What Darwin Got Wrong* (New York: Farrar, Straus and Giroux, 2010), 163.

49. Charles Darwin, *The Origin of Species by Means of Natural Selection, or the Preservation of Favoured Races in the Struggle for Life*, 3rd ed. (New York: Appleton, 1861), chap.14, especially 425.

50. Ronald Hutton, *Queens of the Wild* (New Haven, CT: Yale University Press, 2022), chap. 2, 41–74.

51. On early origins and development of the idea of nature, see George D. Economou, *The Goddess Natura in Medieval Literature*, 2nd ed. (Notre Dame, IN: Notre Dame University Press, 2002), especially 1–27.

52. Hutton, *Queens of the Wild*, 50.

53. Hutton, *Queens of the Wild*, 53.

54. Charles Darwin to Charles Lyell, September 28, 1860, *Darwin Correspondence Project*, Letter no. 2931, University of Cambridge, https://www.darwinproject.ac.uk/letter/?docId=letters/DCP-LETT-2931.xml.

55. Charles Darwin, *The Origin of Species by Means of Natural Selection; Or, the Preservation of Favored Races in the Struggle for Life*, 6th ed. (London: Murray, 1872),

429. The explicit mention of a creator in this concluding passage began with the second edition.

56. G. B. Müller and S. A. Newman, "Origination of Organismal Form: The Forgotten Cause in Evolutionary Theory," in *Origination of Organismal Form: Beyond the Gene in Developmental and Evolutionary Biology*, eds. G. B. Müller and S. A. Newman (Cambridge, MA: MIT Press, 2003), 3–12.

57. Nick Lane, "A Miraculous Merger," in *Chance: The Science and Secrets of Luck, Randomness and Probability*, ed. Michael Brooks (London: Profile/New Scientist, 2015), 22–33.

58. Nick Lane, *The Vital Question: Why Is Life the Way It Is?* (London: Profile, 2015), 2.

59. Lane, "A Miraculous Merger," 22–33, notes 28 and 32.

60. "Surprisingly, natural selection may have little role to play in one of the key steps of evolution—the origin of new species. Instead it would appear that speciation is merely an accident of fate." Bob Holmes, "The Accident of Species," in *Chance: The Science and Secrets of Luck, Randomness and Probability*, ed. Michael Brooks (London: Profile/New Scientist, 2015), 33.

61. Fodor and Piattelli-Palmarini, *What Darwin Got Wrong*, 153, xiv.

62. Darwin, *The Autobiography of Charles Darwin*, 28.

63. Charles Darwin, *Über die Entstehung der Arten*, trans. Heinrich Bronn (Stuttgart: E. Schweizerbart'sche Verlagshandlung und Druckerei, 1860; Charleston, SC: Nabu Press, 2012). Bronn also wrote an appendix for this volume (Schlusswort des Uebersetzers).

64. Bronn's review of the *Origin*, which appeared originally in the *Jahrbuch für Mineralogie* of 1860, 116–120, is reproduced with translation and discussion by David L. Hull in *Darwin and His Critics: The Reception of Darwin's Theory of Evolution by the Scientific Community* (Chicago: University of Chicago Press, 1973), 118–125.

65. Charles Darwin, *Die Entstehung der Arten*, trans. Julius Victor Carus (Stuttgart: Schweitzerbart'sche Verlagsbuchhandlung, 1899; Hamburg, Germany: Nikol, 2018).

66. Eberhard Dennert, *Am Sterbelager des Darwinismus* [1902]. Dennert, *On the Deathbed of Darwinism*, trans. Edwin V. O'Harra and John H. Peschges (Burlington, IA: German Literary Board/R. Neumann, 1904), https://www.gutenberg.org/files/21019/21019-h/21019-h.htm.

67. This expression, used by Hodge when speaking against Darwinism at the 1873 New York meeting of the Evangelical Alliance, is slightly rephrased in his book published the following year, as "blind operation of natural causes, without intention, purpose or cooperation of God." Charles Hodge, *What is Darwinism?* (New York: Scribner, Armstrong and Company, 1874), 64. Cited by Tom McIver, *Anti-Evolution: A Reader's Guide to Writings Before and After Darwin* (Baltimore, MD: Johns Hopkins University Press, 1992), 114.

68. See Neil Broom, *How Blind Is the Watchmaker? Nature's Design and the Limits of Naturalistic Science* (Downers Grove, IL: Intervarsity Press, 2001), 39n.

69. Wallace was fully aware of the necessity of water for life. NASA's seeming obsession with finding evidence of liquid water elsewhere in the solar system suggests that this is still very much an accepted understanding. Alfred Russel Wallace, *The World of Life: A Manifestation of Creative Power, Directive Mind and Ultimate Purpose* (New York: Moffat, Yard and Company, 1911), in particular chap. 18, "The Elements and Water in Relation to the Life-World," 383–397.

70. See Mary Midgely, *Evolution as a Religion* (London: Routledge, 2002) and the various essays edited by Hilary and Steven Rose in *Alas, Poor Darwin: Arguments against Evolutionary Psychology* (London: Vintage, 2001).

71. David Stove, *Darwinian Fairy Tales* (New York: Encounter Books, 2009).

72. John Gray, *Seven Types of Atheism* (London: Penguin, 2019), 14.

73. Antony Flew, *Darwinian Evolution*, 2nd ed. (London: Transaction, 1997).

74. Richard Spilsbury, *Providence Lost: A Critique of Darwinism* (London: Oxford University Press, 1974), 8.

75. Knowing nothing of Mendelian genetics, pangenesis was Darwin's best guess for how heredity operated. He hypothesized (without evidence) that each part of the body released small particles which pooled in the genitals and contributed heritable information to the gametes.

76. Holmes, "The Accident of Species," 34–35.

77. Michael Ruse, *Understanding Natural Selection* (Cambridge, UK: Cambridge University Press, 2023), 133. For a technical discussion of the "statisticalist" point of view, see Charles Pence, *The Causal Structure of Natural Selection* (Cambridge: Cambridge University Press, 2021), 8–11.

CHAPTER 3: PIECING TOGETHER A THEORY

1. Dov Ospovat, *The Development of Darwin's Theory: Natural History, Natural Theology and Natural Selection 1839–1859*, 2nd ed. (Cambridge, UK: Cambridge University Press, 1995), 229–230.

2. See Maud Bodkin, *Archetypal Patterns in Poetry* [1927] (New York: Vintage, 1958).

3. Misia Landau, *Narratives of Human Evolution* (Yale: Yale UP, 1993).

4. Charles Darwin, *Voyage of the Beagle: Charles Darwin's Journal of Researches*, eds. Janet Browne and Michael Neve (London: Penguin, 1989), 2.

5. See John Hedley Brooke, *Science and Religion: Some Historical Perspectives* (Cambridge, UK: Cambridge University Press, 2014), 348–350.

6. Anders Kvernberg, "The Norwegian Who Knew His Tortoises So Well That He Changed the Course of History," *Nypesuppe*, 2017, https://nypesuppe.blogspot .com/2017/06/the-norwegian-who-knew-his-tortoises-so.html. Originally published in Norwegian as *Aftenposten Historie* no. 6 (2015) and *Biolog* no. 1 (2014).

7. Samuel Wilberforce, "On the Origin of Species," *Quarterly Review* (1860): 237.

8. See *Charles Darwin's the Life of Erasmus Darwin*, ed. Desmond King-Hele (Cambridge, UK: Cambridge University Press, 2003), xiii.

9. Charles Lyell, *The Geological Evidence of the Antiquity of Man with an Outline of Glacial and Post-tertiary Geology and Remarks on* The Origin of Species *with Special Reference to Man's First Appearance on the Earth*, 4th ed. (London: John Murray, 1873), 551.

10. Alfred Russel Wallace, "Sir Charles Lyell on Geological Climates and the Origin of Species," *Quarterly Review* 126 (1869): 394.

11. Charles Coulston Gillispie, *Genesis and Geology: The Impact of Scientific Discoveries upon Religious Beliefs in the Decades before Darwin* (New York: Harper, 1959), 126–127.

12. See Charles Lyell, *Principles of Geology: Being an Attempt to Explain the Former Changes of the Earth's Surface, by Reference to Causes Now in Operation*, vol.1 (London: John Murray, 1830–1833), especially chap. 10. This new and more fluid

perception of the world as being gradually but irresistibly on the move and subject to flux impressed itself upon Darwin and was captured well by that poet who frequently exhibited the greatest sensitivity to the scientific zeitgeist, Alfred Lord Tennyson, when he wrote, "The hills are shadows, and they flow / From form to form, and nothing stands / They melt like mist, the solid lands, / Like clouds they shape themselves and go." *In Memoriam* (London: Edward Moxon, 1850), canto 121, https://www.gutenberg.org/cache/epub/70950/pg70950-images.html. Cited by Benjamin Woolley, *Bride of Science: Romance, Reason and Byron's Daughter* (London: Macmillan, 1999), 141.

13. Charles Darwin to Leonard Horner, August 20, 1844, *Darwin Correspondence Project*, Letter no. 771, University of Cambridge, https://www.darwinproject.ac.uk/letter/?docId=letters/DCP-LETT-771.xml. See on this point Loren C. Eiseley, *Darwin and the Mysterious Mr. X* (London: Dent, 1979), 44, https://archive.org/details/darwinmysterious00eise.

14. Sandra Herbert, *Charles Darwin, Geologist* (New York: Cornell University Press, 2005), 356.

15. Charles Darwin, *The Foundations of* The Origins of Species*: Two Essays Written in 1842 and 1844*, ed. Francis Darwin (Cambridge, UK: Cambridge University Press, 1909; London: Lector House, 2020).

16. Charles Darwin, *The Autobiography of Charles Darwin 1809–1882*, ed. Nora Barlow (New York: Norton, 1958), 84.

17. Robert F. Shedinger, *The Mystery of Evolutionary Mechanisms* (Eugene, OR: Cascade, 2019), 40.

18. As Sir Fred Hoyle once commented, the earth's crust shows "no hierarchy of structure, with one level of subtlety piled on another. It is the existence of such a hierarchy which characterizes biological systems. A single crystal of rock, or of a mineral, or of a snowflake, contains subtleties of great interest. But such crystals do not fit together into larger patterns of still greater interest and complexity." Fred Hoyle, *Ten Faces of the Universe* (London: Heinemann, 1977), 164.

19. Barry Gale, *Evolution without Evidence: Charles Darwin and the Origin of Species* (Albuquerque, NM: University of New Mexico Press, 1982), 37.

20. The crowd figure Murellus in Shakespeare's play equates the insensitivity of his peers with the inability of inanimate objects to experience fellow feeling for a departed Roman commander, Pompey. *Julius Caesar*, Act I, scene 1, lines 32–37.

21. Hoyle further elaborates: "Large quantities of inorganic material only *repeat* the simpler forms…. The information content of a blizzard is essentially the same as that of a single snowflake." Hoyle, *Ten Faces of the Universe*, 164.

22. Percival Davis and Dean Kenyon, *Of Pandas and People: The Central Questions of Biological Origins*, 2nd ed. (Dallas, TX: Haughton, 1993), 158–159.

23. Michael Marshall, *The Genesis Quest: The Geniuses and Eccentrics on a Journey to Uncover the Origin of Life on Earth* (London: Weidenfeld and Nicholson, 2020), 11.

24. Carl Sagan titled one of his books *Pale Blue Dot* (New York: Random House, 1994). The descriptor came from the photograph of Earth taken February 14, 1990, by the Voyager 1 space probe from 3.7 billion miles away. Earth appeared as an unremarkable little pale blue dot in the photograph, surrounded by the immensity of space.

25. William Whewell, *Of the Plurality of Worlds* [1853], ed. Michael Ruse (Chicago: University of Chicago Press, 2001).

26. Stephen Craig Dilley, "Methodological Naturalism, History, and Science," abstract (PhD diss., Arizona State University, 2007).

27. Dilley, "Methodological Naturalism," 63. For the Buridan and Oresme quotations, he cites Edward Grant, *God and Reason in the Middle Ages* (Cambridge, UK: Cambridge University Press, 2001), 198 and 199 respectively.

28. Dilley comments, "Bacon thought deeply and wrote extensively about the nature of science and its relationship to religious claims in *Great Instauration*, *Novum Organon*, and *The Advancement of Learning*, works which collectively advocated methodological naturalism." Dilley, "Methodological Naturalism," 2–3.

29. Ronald Numbers, "Science Without God: Natural Laws and Christian Beliefs," in Ronald Numbers, *Science and Christianity in Pulpit and Pew* (New York: Oxford University Press, 2007), 46.

30. Richard Lewontin, *It Ain't Necessarily So: The Dream of the Human Genome and Other Illusions* (London: Granta, 2000), 66.

31. Numbers, "Science Without God," 53.

32. Charles Darwin, *The Autobiography of Charles Darwin 1809–1882*, ed. Nora Barlow (London: Collins, 1958), 120.

33. Francis Darwin, introduction to Charles Darwin, *The Foundation of* The Origin of Species, xvi.

34. See Friedrich Engels to Peter L. Lavrov, November 12, 1875, published in English as "Engels and Darwin—Letter to Lavrov," ed. Dona Torr, *Labour Monthly* (July 1936): 437–442, https://www.marxists.org/archive/marx/works/1875/letters/75_11_12.htm.

35. See on this point Adrian Desmond, *The Politics of Evolution* (Chicago: University of Chicago Press, 1989), 2–3.

36. These included the Oxford Professor of Geometry, Baden Powell; the French naturalist Charles Naudin; Robert Grant (Darwin's Edinburgh tutor); Dr. William Wells (his claim going back to 1813); Patrick Matthew, a well-to-do Scottish farmer and fruit grower (in 1831); and Edward Blyth (1835). A comprehensive account of Darwin's predecessors is given by Rebecca Stott, *Darwin's Ghosts: In Search of the First Evolutionists* (London: Bloomsbury, 2012).

37. Rebecca Stott, *Darwin's Ghosts*, 17.

38. Patrick Matthew, "The Origin of Species," *Gardeners' Chronicle* (May 12, 1860): 433, *Darwin Correspondence Project*, https://darwin-online.org.uk/converted/Ancillary/1860_Matthew_%20A1022.html. For an extended account of Patrick Matthew's contribution to the idea of evolution by natural selection, see W. J. Dempster, *Evolutionary Concepts in the Nineteenth Century: Natural Selection and Patrick Matthew* (Edinburgh: Pentland Press, 1995).

39. Dempster, *Evolutionary Concepts*, 245.

40. Edward Blyth, "An Attempt to Classify the 'Varieties' of Animals with Observations on the Marked Seasonal and Other Changes Which Naturally Take Place in Various British Species, and Which Do Not Constitute Varieties," *Magazine of Natural History* 8, no. 1 (1835), http://www.bradburyac.mistral.co.uk/blyth1.html.

41. Edward Blyth, "On the Psychological Differences between Man and All Other Animals; and the Consequent Diversity of Human Influence over the Inferior Ranks of Creation, from Any Mutual and Reciprocal Influence Exercised Among the Latter," *Magazine of Natural History* (new series) 1, no. 1 (1837): 135, https://archive.org/details/dli.ernet.25788/page/n5/mode/2up. Cited by Eiseley, *Darwin and the Mysterious Mr. X*, 55–56, 58.

42. Here I am paraphrasing Eiseley, *Darwin and the Mysterious Mr. X*, 58. For more on Blyth, including links to transcripts of his original texts, see Salvador Cordova, "Was Blyth the True Scientist and Darwin Merely a Plagiarist and Charlatan?," *Uncommon Descent*, August 26, 2006, https://uncommondescent.com/intelligent-design/was-blyth-the-true-scientist-and-darwin-merely-a-plagiarist-and-charlatan/.

43. Blyth, "On the Psychological Differences" (1837), 135.

44. Blyth, "On the Psychological Differences" (1837), 136.

45. Georges Cuvier, "Extract from a Memoir on an Animal of Which the Bones Are Found in the Plaster Stone… around Paris, and Which Appears No Longer to Exist Alive Today," Read at the public session of the National Institute on… October 6, 1798, in Martin J. S. Rudwick, *Georges Cuvier, Fossil Bones, and Geological Catastrophes: New Translations and Interpretations of the Primary Texts* (Chicago: University of Chicago Press, 1997), 36.

46. Charles Darwin, *The Origin of Species*, 6th ed. (New York: Appleton, 1871), 167–168.

47. For an in-depth treatment of the problem of the missing transitionals, see Stephen C. Meyer, *Darwin's Doubt: The Explosive Origin of Animal Life and the Case for Intelligent Design* (New York: HarperOne, 2013), chaps. 1–4; and *Debating Darwin's Doubt: A Scientific Controversy That Can No Longer Be Denied*, ed. David Klinghoffer (Seattle, WA: Discovery Institute Press, 2015), chaps. 13 and 23 by Casey Luskin, and chap. 14 by Stephen Meyer.

48. Tim Whitmarsh, *Battling the Gods: Atheism in the Ancient World* (London: Faber and Faber, 2016). See also Epicurus, *The Art of Happiness*, trans. George C. Strodach (London: Penguin, 2012), especially 43–45. The notion of a materialistic evolution of nature is of great antiquity, having already been referenced by Heraclitus of Ephesus. See *The First Philosophers: The Presocratics and the Sophists*, trans. Robin Waterfield (Oxford, UK: Oxford University Press, 2000), 32–48.

49. See John W. Judd, *The Coming of Evolution* (Cambridge, UK: Cambridge University Press, 1911), 3.

50. Epicurus's philosophical speculations were entirely material. He conceived of nature as, and only as, an aggregate of material entities operated by blind and unvarying natural laws, without enquiring as to the source of such laws.

51. See David Hume, *Dialogues and Natural History of Religion*, ed. J. C. A. Gaskin (Oxford, UK: Oxford University Press, 2008).

52. Stott, *Darwin's Ghosts*, 40.

53. See the discussion of David Sedley, *Creationism and its Critics in Antiquity* (Berkeley, CA: University of California Press, 2009), 155–156.

54. Cicero, *The Nature of the Gods*, trans. P. G. Walsh (New York: Oxford University Press, 1998), 80–81. David Sedley discussed Cicero and his argument in an interview with Catherine Wilson, Bridget Kendall, and Sonia Wurster on BBC's *The Forum*, "In Search of the Good Life: Epicurus and His Philosophy," April 2, 2020, audio, 44:00, https://www.bbc.co.uk/sounds/play/w3cszjv4. The relevant section begins at approximately the 12-minute mark.

55. Fred Hoyle, *The Intelligent Universe* (London: Michael Joseph, 1983), 11–12.

56. See Fergus Kerr, *Thomas Aquinas: A Very Short Introduction* (Oxford, UK: Oxford University Press, 2009), 39–40.

57. See Basil Willey, *The Seventeenth-Century Background: Studies in the Thought of the Age in Relation to Poetry and Religion* (London: Pelican, 1972), 111–122.

58. Sedley, *Creationism and its Critics*, 134.

59. Somewhat curiously, Epicurus went along with the then-universal consensus that the gods existed, but he held that they were wholly impassible and morally disengaged and therefore pragmatically irrelevant to humankind and its concerns.

60. See Whitmarsh, *Battling the Gods*, 173–185.

61. As Oxford mathematics professor John Lennox noted, "I do not associate the existence of God as much with worry but with joy." See *Gunning for God: Why the New Atheists Are Missing the Target* (Oxford, UK: Lion Hudson, 2011), 12.

62. Sedley, *Creationism and Its Critics*, 242.

CHAPTER 4: VICTORIAN CULTURE WARS

1. Mrs. Humphry Ward, introduction to *The Writings of Mrs. Humphry Ward: Robert Elsmere* (New York: Houghton Mifflin, 1909), xxii. Cited by Michael Ruse, *Darwinism as Religion: What Literature Tells Us about Evolution* (Oxford, UK: Oxford University Press, 2017), x.

2. For a full account of Swinburne's militantly turning his back on the Christian faith, see Bernard Schweizer, *Hating God: The Untold Story of Misotheism* (Oxford, UK: Oxford University Press, 2011), 83–101.

3. Alec Ryrie, *Unbelievers: An Emotional History of Doubt* (London: Collins, 2019), 4n151.

4. Ruse, *Darwinism as Religion*, 7.

5. Schweizer, *Hating God*, 5.

6. The phrase is that of A. N. Wilson, *God's Funeral* (London: John Murray, 1999), 205.

7. See Schweizer, *Hating God*, 83–101. Also, Kathleen Margot Louis, *Swinburne and His Gods: The Roots and Growth of Agnostic Poetry* (Montreal: McGill University Press, 1990).

8. Algernon Charles Swinburne, "Hymn to Proserpine (After the Proclamation in Rome of the Christian Faith)," line 32, https://www.poetryfoundation.org/poems/45292/hymn-to-proserpine-after-the-proclamation-in-rome-of-the-christian-faith.

9. Schweizer, *Hating God*, 92.

10. On Swinburne's hell-raising see Philip Henderson, *Swinburne: The Portrait of a Poet* (London: Routledge and Kegan Paul, 1974).

11. The analogy is that of A. N. Wilson, *God's Funeral*, 210.

12. This anecdote is mentioned in A. N. Wilson, *God's Funeral*, in his chapter "Swinburne and the Gods," 203–230.

13. Charles Darwin, *The Autobiography of Charles Darwin 1809–1882*, ed. Nora Barlow (New York: Norton, 1958), 85–96.

14. Charles Darwin to Asa Gray, May 22, 1860, *Darwin Correspondence Project*, Letter no. 2855, University of Cambridge, https://www.darwinproject.ac.uk/letter/?docId=letters/DCP-LETT-2814.xml.

15. I am not suggesting that an unmuddled and profitable cross-fertilization between scientific and theological thinking isn't possible. Indeed, the scientific revolution is

widely understood, by secular and religious historians alike, to have been to a substantial degree the product of Judeo-Christian thinking about the created order and mankind's relationship to that order and its Creator. See Stephen Dilley, "Darwin's Use of Theology in the *Origin of Species*," *British Journal for the History of Science* 45, no.1 (2012): 29–57; and by the same author, "The Evolution of Methodological Naturalism in *The Origin of Species*," *British Journal for the History of Science* 44, no. 1 (2012): 29–56.

16. Simon Powell, *Darwin's Unfinished Business: The Self-Organizing Intelligence of Nature* (Rochester, VT: Park Street Press, 2012), 18.

17. William Irvine, *Apes, Angels and Victorians: A Joint Biography of Darwin and Huxley* (London: Weidenfeld and Nicholson, 1955), 71.

18. Some objected that present developments would lead to Oxford's becoming an irreligious institution (a not unwarranted fear, in the event).

19. See Adrian Desmond, *Huxley: From Devil's Disciple to Evolution's High Priest* (London: Penguin, 1997), 385.

20. Owen Chadwick, *The Secularization of the European Mind in the Nineteenth Century* (Cambridge, UK: Cambridge University Press, 1975), 27.

21. For information on this subject I am indebted to Josef F. Altholz, *Anatomy of a Controversy: The Debate over "Essays and Reviews"* (Aldershot, UK: Scolar Press, 1994). The contributors to *Essays and Reviews* were Frederick Temple, Rowland Williams, Baden Powell, Henry Bristow Wilson, C. W. Goodwin, Mark Pattison, and Benjamin Jowett.

22. Samuel Taylor Coleridge, *Confessions of an Enquiring Spirit* (London: Cassell and Company, 1892), Letter 4, https://www.gutenberg.org/ebooks/2575. Thomas Arnold, headmaster of Rugby School, was another early proponent of what became known known as the historical-critical approach to the Bible.

23. Benjamin Jowett, "On the Interpretation of Scripture," in *Essays and Reviews* [1869] (London: Longman, Green, Longman, and Roberts, 1862), 409, https://archive.org/details/dli.ministry.12495/page/n5/mode/2up.

24. Gotthold Ephraim Lessing wrote *"Der Buchstabe ist nicht der Geist,"* or "The letter is not the spirit" in *Ein Mehreres aus den Papieren des Ungenannten, die Offenbarung betreffend* [1777]. German eighteenth- and early nineteenth-century biblical scholarship was influenced in some part by the English deists of the seventeenth century.

25. Altholz, *Anatomy of a Controversy*, 33.

26. Altholz, *Anatomy of a Controversy*, 37.

27. Altholz, *Anatomy of a Controversy*, 44.

28. For Matthew Tindall, in his *Christianity as Old as Creation* (1730), Christianity meant little more than good citizenship. He pointed out that many had lived before the time of Christ, so how could such ancestors be "justified"? Tindall's answer: by their reason, not by faith.

29. Mark Pattison, "Tendencies of Religious Thought in England, 1688–1750," in *Essays and Reviews*, 310.

30. For details see Edward Royal, *The Infidel Tradition:From Paine to Bradlaugh* (London: MacMillan, 1976) and, for a detailed contextualization, Adrian Desmond, *The Politics of Evolution: Morphology, Medicine and Reform in Radical London* (Chicago: Chicago University Press, 1992).

31. Timothy Larsen, *Crisis of Doubt: Honest Faith in Nineteenth-Century England* (Oxford, UK: Oxford University Press, 2006), 82–83.

32. See Noel Annan, *Leslie Stephen, the Godless Victorian* (New York: Random House, 1984), 45. See also Larsen, *Crisis of Doubt*, 247–248.

33. Larsen, *Crisis of Doubt*, 248.

34. Jennifer Hecht, *Doubt: A History* (New York: HarperCollins, 2004), 407.

35. On Darwin's distinctly awkward dealings with these two militant secularists, Charles Bradlaugh and Edward Aveling, see James Moore, *The Darwin Legend* (London: Hodder and Stoughton, 1994), 23–29. The reason for the rejection may have been that Darwin did not want to be connected with anyone who might prove politically radical. Here, noted Moore, was "the parish naturalist seeking not to disturb the social equilibrium." Darwin's anxieties were not entirely misplaced, for grassroots secularism had been closely linked with political radicalism ever since Thomas Paine's *The Rights of Man* (1791–1792) had laced ideals of political emancipation with critiques of biblical anomalies worthy of the German Higher Criticism. Paine, later to be driven into exile, was widely seen as a public enemy who had defended not only the French Revolution but also the American Revolution, the latter of which had deprived Britain of its most valued overseas colony. The bourgeois Darwin was then understandably wary of having his name linked with persons potentially capable of political insurrection.

36. William Shakespeare, *Othello*, Act 3, scene 2, lines 373–376.

37. Larsen, *Crisis of Doubt*, 242.

38. Larsen, *Crisis of Doubt*, 243.

CHAPTER 5: DISSENTING VOICES

1. For full accounts of Wallace's career, see Ross Slotten, *The Heretic in Darwin's Court: The Life of Alfred Wallace* (New York: Columbia University Press, 2004); Michael Flannery, *Alfred Russell Wallace: A Rediscovered Life* (Seattle, WA: Discovery Institute Press, 2011) and *Nature's Prophet: Alfred Russel Wallace and his Evolution from Natural Selection to Natural Theology* (Tuscaloosa, AL: University of Alabama Press, 2018).

2. Caroline Catchpole, "I Am Afraid the Ship's on Fire," *Alfred Russel Wallace Correspondence Project*, April 30, 2014, https://wallaceletters.myspecies.info/content/%E2%80%9Ci-am-afraid-ship%E2%80%99s-fire%E2%80%9D.

3. Alfred Russel Wallace to Richard Spruce, September 19, 1852, *Alfred Russel Wallace Correspondence Project*, Letter no. WCP349.349, https://epsilon.ac.uk/view/wallace/letters/WCP349#back-mark-LWCP349-349-foot.f1.

4. Robert Chambers, *Vestiges of the Natural History of Creation* (London: John Churchill, 1844), https://www.gutenberg.org/cache/epub/7116/pg7116.txt.

5. Chambers, *Vestiges*, 146.

6. Chambers, *Vestiges*, li. Chambers further assured readers that even without God's hand in direct creation, "we may still feel that He is the immediate breather of our life and ruler of our spirits," 323.

7. Chambers, *Vestiges*.

8. On the subject, see Flannery, *Nature's Prophet*, 18–21.

9. Daniel Stashower, *Teller of Tales: The Life of Sir Arthur Conan Doyle* (New York: Henry Holt, 1999), 93.

10. Stashower, *Teller of Tales*, 162.

11. Alfred Russel Wallace, *Man's Place in the Universe* [1903] (London: Amazon Publishing, 2015).

12. Alfred Russel Wallace, *The World of Life* [1910]; reprinted as *Intelligent Evolution: How Alfred Russel Wallace's World of Life Challenged Darwinism*, ed. Michael Flannery (Nashville: Erasmus Press, 2020).

13. Alfred Russel Wallace, *Is Mars Habitable?* [1907] (London: Amazon CreateSpace, 2016).

14. Alfred Russel Wallace, "The Origin of Human Races and the Antiquity of Man Deduced from the Theory of 'Natural Selection,'" *Journal of the Anthropological Society of London* 2 (1864): 158–187, https://wallace-online.org/content /frameset?itemID=S093&viewtype=text&pageseq=9.

15. Alfred Russel Wallace, "Sir Charles Lyell on Geological Climates and the Origin of Species," *Quarterly Review* 126 (1869): 394.

16. Charles Darwin to A. R. Wallace, March 27, 1869, *Darwin Correspondence Project*, Letter no. 6684, University of Cambridge, https://www.darwinproject.ac.uk /letter/?docId=letters/DCP-LETT-6684.xml.

17. P. S., "Wallace on Natural Selection" [review of Alfred Russel Wallace's *Contributions to the Theory of Natural Selection*], *Nature* 2 (October 13, 1870): 472.

18. Michael A. Flannery, "Alfred Russel Wallace's Intelligent Evolution and Natural Theology," *Religions* 11 (2020): 12, https://www.mdpi.com/2077-1444/11/6/316.

19. Rebecca Stott, *Darwin's Ghosts: In Search of the First Evolutionists* (London: Bloomsbury, 2012), 276.

20. See Paul Harrison, *Elements of Pantheism*, 3rd ed. (Shaftesbury, UK: Element, 2013). Otto's book *Das Heilige* appeared in 1917 (in English as *The Idea of the Holy* in 1923).

21. Alfred Russel Wallace, *Darwinism: An Exposition of the Theory of Natural Selection* [1889] (London: Macmillan, 1890), https://archive.org/details/in.ernet.dli.2015 .221825/page/n491/mode/2up.

22. Wallace, *Darwinism*, 474–475.

23. Frank M. Turner, *Between Science and Religion: The Reaction to Scientific Naturalism in Late Victorian England* (New Haven, CT; Yale University Press, 1974), 72–73.

24. Oliver Goldsmith, *A Good-Natur'd Man: A Comedy*, in *Poems and Plays: By Oliver Goldsmith, M. B. to which is Prefixed, the Life of the Author* (Dublin: William Wilson, 1777), https://name.umdl.umich.edu/004771299.0001.000.

25. Alfred Russel Wallace to James Marchant, April 26, 1913, *Alfred Russel Wallace Correspondence Project*, Letter no. WCP6558.7574, https://epsilon.ac.uk/view /wallace/letters/WCP6568.

26. Charles Darwin to Charles Lyell, September 28, 1860, *Darwin Correspondence Project*, Letter no. 2931, University of Cambridge, https://www.darwinproject. ac.uk/letter/?docId=letters/DCP-LETT-2931.xml.

27. Charles Darwin, *The Descent of Man, and Selection in Relation to Sex* [1871], ed. James Moore and Adrian Desmond (London: Penguin, 2004), 110.

28. A giraffe for instance cannot elongate its neck (and hence the necks of its progeny) by repeatedly craning towards the higher branches of trees.

29. Klaus Taschwer, *The Case of Paul Kammerer: The Most Controversial Biologist of His Time* (Montreal, Canada: Bunim and Bannigan, 2019), 9.

30. See Taschwer, *The Case of Paul Kammerer*, chap. 1.

31. See on this point Arthur Koestler, *The Case of the Midwife Toad* (London: Hutchinson, 1971), 27–30.

32. Paul K. Kammerer, *The Inheritance of Acquired Characteristics* (New York: Boni and Liveright, 1924), 261–65.

33. Koestler cited with approval the view of the mid-twentieth-century scientist C. H. Waddington that chance mutation was like throwing bricks together in heaps in the hope that they would arrange themselves into an inhabitable house.

34. Koestler, *The Case of the Midwife Toad*, 129. For a balanced assessment of Koestler's intellectual achievements and weaknesses, see Michael Scammell, *Koestler: The Indispensable Intellectual* (London: Faber and Faber, 2011).

35. Ludwig von Bertalanffy, "Chance or Law," in *Beyond Reductionism: The Alpbach Symposium 1968*, eds. Arthur Koestler and J. R. Smythies (London: Hutchinson, 1969), 67.

36. See John and Mary Gribbin, *On the Origin of Evolution: Tracing Darwin's Dangerous Idea from Aristotle to DNA* (London: Collins, 2020), 230–252; Paul Davies, *The Demon in the Machine: How Hidden Webs of Information Are Solving the Mystery of Life* (London: Penguin, 2020), 109–143; and Nessa Carey, *The Epigenetics Revolution: How Modern Biology Is Rewriting Our Understanding of Genetics, Disease and Inheritance* (London: Icon, 2011).

37. See the conspectus of diverging modern views covered by Stephen Buranyi in "Do We Need a New Theory of Evolution?," *Guardian*, June 28, 2022.

38. Gillian Beer, *Darwin's Plots: Evolutionary Narrative in Darwin, George Eliot and Nineteenth-Century Fiction* (London: Routledge, 1983), 9.

39. *Beseelt*, from German *Seele* (soul). The *be*-prefix has transitive force, hence "endowed with soul," in older English "ensouled."

40. Alfred Kelly, *The Descent of Darwin: The Popularization of Darwinism in Germany 1860–1914* (Chapel Hill, NC: North Carolina University Press, 1981), 28.

41. Kelly, *Descent of Darwin*, 84.

42. See J. David Pleins, *In Praise of Darwin: George Romanes and the Evolution of a Darwinian Believer* (New York: Bloomsbury, 2014), 299–348.

43. John Bowlby gives a useful Who's Who of the characters who played a part in Darwin's career in *Charles Darwin: A New Biography* (London: Pimlico, 1991), 489–504.

44. For Henslow and his relationship with Darwin, see Jean Russell-Gebbet, *Henslow of Hitcham* (Lavenham, UK; Terence Dalton, 1977).

45. Russell-Gebbet, *Henslow of Hitchen*, 23–24.

46. Chris Park, *Wedded to the Rocks: The Life and Work of Adam Sedgwick* [1785–1873] (London: Amazon CreateSpace, 2017), 244.

47. Iain McCalman, *Darwin's Armada: Four Voyages and the Battle for the Theory of Evolution* (London: Norton, 2009), 315.

48. The term public school is a dishonest misnomer. It means *private* school in the British context.

49. See Edward Lurie, *Louis Agassiz: A Life in Science* [1960] (Baltimore, MD: Johns Hopkins University Press, 1988), 287.

50. A. Hunter Dupree, *Asa Gray: American Botanist, Friend of Darwin* [1959] (Baltimore, MA: Johns Hopkins University Press, 1988), especially 269–271.

51. Lurie, *Louis Agassiz*, 255.

52. Lurie, *Louis Agassiz*, 298.

53. Peter Bowler, *The Non-Darwinian Revolution* (Baltimore: Johns Hopkins University Press, 1988), 68.

54. David Orenstein and Abby Hafer, *Darwin's Apostles: The Men Who Fought to Have Evolution Accepted, Their Times, and How the Battle Continues* (Washington DC: Humanities Press, 2019), 6.

55. In 1859 Huxley published "Mr. Darwin's *Origin of Species*" in *MacMillan's Magazine*. In December of the same year he published a very favorable review of the *Origin* in *The Saturday Review*. In 1863 he published the Darwin-friendly *Man's Place in Nature* and in 1864 he founded the X-Club. In 1871 there appeared his *Mr. Darwin's Critics*, a collection of essays defending Darwin from his detractors. And as late as 1880 he addressed the Royal Society on the subject of "The Coming of Age of *The Origin of Species*."

CHAPTER 6: NATURE'S GOD

1. "This conception of the world we call mythological because it is different from the conception of the world which has been formed and developed by science since its inception in ancient Greece and which has been accepted by all modern men. In this modern conception of the world the cause-and-effect nexus is fundamental. Although modern physical theories take account of chance in the chain of cause and effect in subatomic phenomena, our daily living, purposes and actions are not affected. In any case, modern science does not believe that the course of nature can be interrupted or, so to speak, perforated, by supernatural powers." Rudolf Bultmann, *Jesus Christ and Mythology* (London: SCM Press, 1960), 15.

2. See for example *St. Augustine: The Literal Meaning of Genesis*, trans. John Hammond Taylor (New York: Paulist Press, 1982). For a broader analysis of the views of the Early Fathers, see Craig D. Allert, *Early Christian Readings of Genesis One: Patristic Exegesis and Literal Interpretation* (Downers Grove, IL: IVP Academic, 2018) or David Knight, *Science and Spirituality: The Volatile Connection* (London: Routledge, 2004).

3. See Karen Armstrong, *The Bible: A Biography* (London: Atlantic Books, 2007), 47–48.

4. See Michael Ruse, *The Gaia Hypothesis: Science on a Pagan Planet* (Chicago: Chicago University Press, 2013).

5. For further discussion of this point see Sir Edward Bailey's *Charles Lyell* (London: Thomas Nelson, 1962), 17.

6. Darwin's view that the marvelous elements of the Bible are mere myth is decipherable from the jocular tone of his PS remark in his letter to George Warington, October 7, 1867, *Darwin Correspondence Project*, Letter no. 5642, University of Cambridge, https://www.darwinproject.ac.uk/letter/?docId=letters/DCP-LETT-5642.xml. That outlook is made explicit elsewhere as, for example, Darwin's letter to John Crawfurd, April 7, 1861, *Darwin Correspondence Project*, Letter no. 3114, University of Cambridge, https://www.darwinproject.ac.uk/letter/?docId=letters/DCP-LETT-3114.xml; and in Darwin's brief letter to Frederick McDermott, November 24, 1880, where he bluntly states that "I do not believe in the Bible as divine revelation." *Darwin Correspondence Project*, Letter no.

12851, University of Cambridge, https://www.darwinproject.ac.uk/letter/?docId
=letters/DCP-LETT-12851.xml.

7. Charles Darwin, *The Autobiography of Charles Darwin 1809–1882*, ed. Nora Barlow (New York: Norton, 1958), 85.

8. Nick Spencer, *Darwin and God* (London: SPCK, 2009), 43.

9. Robert Reiss, *Sceptical Christianity: Exploring Credible Belief* (London: Jessica Kingsley, 2016), 29.

10. Already in the previous century the German dramatist Lessing had opposed biblical literalism and drawn a firm distinction between the letter and spirit of the Bible ("Der Buchstabe ist nicht der Geist"). See Henry Chadwick, *Lessing's Theological Writings* (Stanford, CA: Stanford University Press, 1957).

11. I refer here to such figures in the British context as Bishop John Robinson in his *Honest to God* from the 1960s or, some decades later, to Bishop David Jenkins in his Bultmann-inspired rejection of precise dogmas and miracles such as the virgin birth, the resurrection, and walking on water, all of which he glossed as the exuberant result of Jesus's early followers' faith. See the essays in *Honest to God: Forty Years On*, ed. Colin Slee (London: SCM, 2004).

12. Northrop Frye, *The Great Code: The Bible and Literature* (New York: Harcourt Brace Jovanovich 1982)

13. Psalm 14:1.

14. See Timothy Larsen, *A People of the Book: The Bible and the Victorians* (Oxford, UK: Oxford University Press, 2012), 70, 195–196.

15. William James's attitude to organized religion was ambivalent since he believed that special manifestations of religion with their formal creeds and associated shibboleths might sometimes be absurd, yet he still felt that religion had "an important function" for mankind. See Martin E. Marty, introduction to William James, *The Varieties of Religious Experience* (New York: Penguin Classics, 1982), xix.

16. Emma Darwin to Charles Darwin, February 1839, *Darwin Correspondence Project*, Letter no. 471, University of Cambridge, https://www.darwinproject.ac.uk/letter/?docId=letters/DCP-LETT-471.xml. The letter is also published in vol. 2 of *The Correspondence of Charles Darwin*, eds. Frederick Burkhardt et al. (Cambridge, UK: Cambridge University Press, 1985-) and in Edna Healey, *The Inspirational Wife of a Genius* (London: Headline Book Publishing, 2001), 163–164.

17. Alvar Ellegard, *Darwin and the General Reader: The Reception of Darwin's Theory of Evolution in the British Periodical Press 1859–1872* [1958] (Chicago: University of Chicago Press, 1990).

18. Darwin, *The Autobiography of Charles Darwin*, 311–312.

19. Darwin, *The Autobiography of Charles Darwin*, 138–139.

20. For biographical details see Stephen Gill, *Wordsworth: A Life*, 2nd ed. (Oxford, UK: Oxford University Press, 2020).

21. J. R. Watson, *Wordsworth's Vital Soul: The Sacred and Profane in Wordsworth's Poetry* (London: Macmillan, 1982), 2.

22. William Wordsworth, "Lines Composed a Few Miles above Tintern Abbey," July 13, 1798, https://www.poetryfoundation.org/poems/45527/lines-composed-a-few-miles-above-tintern-abbey-on-revisiting-the-banks-of-the-wye-during-a-tour-july-13-1798.

23. William Hale White, *The Autobiography of Mark Rutherford, Dissenting Minister* [1881], ed. William S. Peterson (Oxford, UK: Oxford University Press, 1990).

24. White, *Autobiography of Mark Rutherford*, 21–22.

25. Percy Bysshe Shelley, *A Defence of Poetry*, ed. Mary Shelley [1840] (Indianapolis, IN: Bobbs-Merrill, 1904), 90, https://archive.org/details/adefencepoetry 00shelgoog/page/n96/mode/2up?q=hierophants. Shelley wrote this essay in 1821 and it was published posthumously.

26. For an account of Wordsworth's later influence, see Jonathan Bate, *Radical Wordsworth: The Poet Who Changed the World* (London: HarperCollins, 2020), 449–479. The most exhaustive account of reception is provided in Robert M. Ryan's *Charles Darwin and the Church of Wordsworth* (Oxford, UK: Oxford University Press, 2016).

27. Stopford Brooke, *Theology in the English Poets: Cowper, Coleridge, Wordsworth and Burns* [1872] (London: Dent, 1910), 79.

28. Frederic William Henry Myers, *Wordsworth* [1881] (Hamburg: Tredition, 2012), 113.

29. Myers, *Wordsworth*, 118.

30. Bate, *Radical Wordsworth*, 464.

31. For a discussion of Wordsworth's effect on James, see David E. Leary, "'Authentic Tidings': What Wordsworth Gave to William James," *William James Studies* 13, no. 1 (Spring 2017), https://scholarship.richmond.edu/cgi/viewcontent. cgi?params=/context/psychology-faculty-publications/article/1061/&path_info =Authentic_Tidings_What_Wordsworth_gave_to_William_James.pdf.

32. Matthew Arnold, *Essays in Criticism, Second Series* [1888] (London: MacMillan, 1913), 2–3, https://archive.org/details/cu31924012543058/page/n17/mode/2up?q =%22breath+and+finer+spirit%22.

33. Rudolf Otto, *The Idea of the Holy*, trans. J. Harvey (Oxford, UK: Oxford University Press, 1958).

34. Mircea Eliade, *The Sacred and the Profane: The Nature of Religion* [1957] (New York: Harcourt, 1968), 21.

35. Bate, *Radical Wordsworth*, 479.

36. See Ryan, *Charles Darwin and the Church of Wordsworth*, 85n252.

37. Karen Armstrong, *Sacred Nature: How We Can Recover Our Bond with the Natural World* (London: Vintage, 2023), 77.

38. Bate, *Radical Wordsworth*, 483.

39. See Neil Thomas, *Taking Leave of Darwin: A Longtime Agnostic Discovers the Case for Design* (Seattle, WA: Discovery Institute Press, 2021), 33–37.

40. Ryan, *Charles Darwin and the Church of Wordsworth*, 80.

CHAPTER 7: DARWINISM AND THE LOSS OF THE ENLIGHTENMENT PARADIGM

1. Jennifer Hecht, *Doubt: A History* (New York: HarperOne, 2003), 402.

2. M. H. Abrams, *Natural Supernaturalism: Tradition and Revolution in Romantic Literature* (New York: Norton, 1971), 64.

3. Roy Porter, *Enlightenment: Britain and the Creation of the Modern World* (London: Penguin, 2001), 445.

4. See on this point Marcus du Sautoy, *What We Cannot Know: From Consciousness to the Cosmos* (London: Fourth Estate, 2017) and Carlo Rovelli, *Reality Is Not What It Seems: The Journey to Quantum Gravity* (London: Penguin, 2016).

5. Isaac Newton, *General Scholium to the Philosophiæ Naturalis Principia Mathematica*, 3rd ed. [1726], trans. Andrew Motte, 1729, https://isaacnewton.ca/wp-content/uploads/2013/06/newton-general-scholium-1729-english-text-by-motte-letter-size.pdf. In that same text Newton is also at pains to differentiate his view from pantheism, with its view of the universe as something like the soul of God.

6. J. B. S. Haldane, *Possible Worlds and Other Essays* (London: Chatto and Windus, 1927), 286.

7. Kenneth R. Miller, *Finding Darwin's God: A Scientist's Search for Common Ground Between God and Evolution* (New York: Harper Collins, 2008), 219.

8. Steve Taylor, *Spiritual Science* (London: Watkins, 2018), 215.

9. Christian Bandea, *God of the Details* (London: Amazon, 2021), 40.

10. Richard Dawkins, Sam Harris, Daniel Dennett, and Christopher Hitchens, *The Four Horsemen: The Discussion that Caused an Atheist Revolution*, ed. Stephen Fry (London: Bantam, 2019).

11. Peter Vorzimmer, *Darwin: The Years of Controversy* (London: University of London Press, 1972), 254.

12. See Stephen C. Meyer, *Return of the God Hypothesis: Three Scientific Discoveries That Reveal the Mind Behind the Universe* (New York: HarperOne, 2021), 396–406.

13. Charles Townes, interviewed by Bonnie Azab Powell, "'Explore as Much as We Can': Nobel Prize Winner Charles Townes on Evolution, Intelligent Design, and the Meaning of Life," *UC Berkeley News*, June 17, 2005, http://www.freerepublic.com/focus/f-news/1834439/posts?page=158.

14. See for instance Paul Davies's *The Mind of God: Science and the Search for Ultimate Meaning* (London: Penguin, 1992), 15–16.

15. William Paley, *Natural Theology* [1802], eds. Matthew D. Eddy and David Knight (Oxford, UK: Oxford University Press, 2008), 7–8.

16. Cicero, *On the Nature of the Gods*, trans. Horace McGregor (Harmondsworth, UK: Penguin, 1972), 158–159.

ACKNOWLEDGMENTS

I would like to thank the Discovery Institute Press team for reposing their trust in me and for their willingness to publish what is my second volume with them. In particular I would like to thank Jonathan Witt for his truly indefatigable labors in knocking the manuscript into its final shape. Thanks are also due to David Klinghoffer, the editor of the electronic resource *Evolution News & Science Today*, where I was able to publish early forms of some of the ideas which I have now further developed in book form. In a less formal vein I enjoyed the opportunity of discussing matters of mutual interest with Steve Laufmann and his wife in Reading in late summer 2024. Last but by no means least, I would like to thank my wife, Professor Emerita Francoise Le Saux for helping me so skillfully and so selflessly throughout the long gestation of the now happily finished product.

INDEX